RESEARCH ETHICS FORUM

Volume 4

Editors in Chief:

David Hunter, *University of Birmingham, UK*
John McMillan, *University of Otago, Canada*
Charles Weijer, *The University of Western Ontario, Canada*

EDITORIAL BOARD:

Godfrey B. Tangwa
University of Yaounde, Cameroon

Andrew Moore
University of Otago, New Zealand

Jing-Bao Nie
University of Otago, New Zealand

Ana Borovečki
University of Zagreb, Croatia

Sarah Edwards
University College London, UK

Heike Felzmann
National University of Ireland, Ireland

Annette Rid
University of Zurich, Switzerland

Mark Sheehan
University of Oxford, UK

Robert Levine
Yale University, USA

Alex London
Carnegie Mellon University, USA

Johnathan Kimmelman
McGill University, Canada

The series, Research Ethics Forum aims to encourage discussion in the field of research ethics and the ethics of research. Volumes included can range from foundational issues to practical issues in research ethics. No disciplinary lines or borders are drawn and submissions are welcome from all disciplines as well as scholars from around the world. We are particularly interested in texts addressing neglected topics in research ethics, as well as those which challenge common practices and beliefs about research ethics. By means of this Series we aim to contribute to the ever important dialogue concerning the ethics in how research is conducted nationally and internationally. Possible topics include: Research Ethics Committees, Clinical trials, International research ethics regulations, Informed consent, Risk-benefit calculations, Conflicts of interest, Industry funded research, Exploitation, Qualitative research ethics, Social science research ethics, Ghostwriting, Bias, Animal research, Research participants.

More information about this series at http://www.springer.com/series/10602

Daniel Strech • Marcel Mertz
Editors

Ethics and Governance of Biomedical Research

Theory and Practice

Editors
Daniel Strech
Institute for History, Ethics and Philosophy
  of Medicine
Hannover Medical School
Hannover, Germany

Marcel Mertz
Institute for History, Ethics and Philosophy
  of Medicine
Hannover Medical School
Hannover, Germany

Research Unit Ethics, Institute for History
  and Ethics of Medicine
University Hospital Cologne
Cologne, Germany

ISSN 2212-9529   ISSN 2212-9537   (electronic)
Research Ethics Forum
ISBN 978-3-319-28729-4   ISBN 978-3-319-28731-7   (eBook)
DOI 10.1007/978-3-319-28731-7

Library of Congress Control Number: 2016939956

© Springer International Publishing Switzerland 2016
This work is subject to copyright. All rights are reserved by the Publisher, whether the whole or part of the material is concerned, specifically the rights of translation, reprinting, reuse of illustrations, recitation, broadcasting, reproduction on microfilms or in any other physical way, and transmission or information storage and retrieval, electronic adaptation, computer software, or by similar or dissimilar methodology now known or hereafter developed.
The use of general descriptive names, registered names, trademarks, service marks, etc. in this publication does not imply, even in the absence of a specific statement, that such names are exempt from the relevant protective laws and regulations and therefore free for general use.
The publisher, the authors and the editors are safe to assume that the advice and information in this book are believed to be true and accurate at the date of publication. Neither the publisher nor the authors or the editors give a warranty, express or implied, with respect to the material contained herein or for any errors or omissions that may have been made.

Printed on acid-free paper

This Springer imprint is published by Springer Nature
The registered company is Springer International Publishing AG Switzerland

Preface

This edited volume is the result of a 5 day conference at Hannover Medical School in August 2013, focussing on current challenges in preclinical, clinical, and public health research from a research ethics and governance perspective. Experienced young scholars (postdoctoral researchers or PhD candidates) from different disciplines—medicine, philosophy, biology, public health, jurisprudence, and others—were invited to present their recent theoretical or empirical research in the field, and to critically discuss their findings with several leading experts, who also held separate workshops about current challenges in research governance.

We were delighted that most of the participants were willing to put additional work into their valuable contributions and to write the manuscripts that are part of this edited volume. Each chapter in this volume was peer-reviewed by another contributing author, as well as by the editors. Therefore, we are indebted to all contributing authors, who have not only put effort into their texts to make this edited volume possible, but who were also ready to review another authors' chapter and revise their own chapters based on its respective review, thus contributing to the quality of this book.

The chapters in this volume address a wide range of complex current challenges in biomedical research ethics and often propose new ways to handle them. The volume is valuable for all researchers in either research ethics/governance or biomedical research itself, in advancing academic debate and providing new perspectives or concepts for reflecting actual practice. However, besides being of interest to experts on the topic, the volume may also function as a source of supplementary reading for study courses in disciplines such as medicine, public health, applied ethics, philosophy, law, or social sciences.

We would like to express our appreciation to all participants as well as to the supporting staff (Britta Sander, Hannes Kahrass, and Irene Hirschberg) of the conference in Hannover; without their contributions, discussions and constructive critique, it would not have been the successful event that it was.

We would also like to thank Springer International for making this volume possible. Additional thanks go to the following people who undertook editing (Jan Schürmann) and proof reading (Anja Löbert/Textwork Solutions, Reuben Thomas).

Finally, the German Federal Ministry of Education and Research (Bundesministerium für Bildung und Forschung, BMBF) deserves our thanks for providing generous funding, which enabled us to attract national and international experts in the field to our conference in Hannover and to organise this edited volume.

Hannover, Germany
February 2016

Daniel Strech
Marcel Mertz

Contents

1   **Introduction** 1
Daniel Strech and Marcel Mertz

Part I   Introducing New Domains of Research Governance

2   **Should Research Ethics Encourage the Production
of Cost-Effective Interventions?** 13
Govind Persad

3   **From Altruists to Workers: What Claims Should Healthy
Participants in Phase I Trials Have Against Trial Employers?** 29
Rebecca A. Johnson

4   **Nocebo Effects: The Dilemma of Disclosing Adverse Events** 47
Luana Colloca

Part II   Challenges in Common Domains of Research Governance

5   **Discriminating Between Research and Care in Paediatric
Oncology—Ethical Appraisal of the ALL-10 and 11 Protocols
of the Dutch Childhood Oncology Group (DCOG)** 59
Sara A.S. Dekking, Rieke van der Graaf, Martine C. de Vries,
Marc B. Bierings, and Johannes J.M. van Delden

6   **What Does the Child's Assent to Research
Participation Mean to Parents? Empirical Findings
in Paediatric Oncology in Germany** 73
Imme Petersen and Regine Kollek

7   **Assent in Paediatric Research and Its Consequences** 87
Jan Piasecki, Marcin Waligora, and Vilius Dranseika

8   **Ethical Principles in Phase IV Studies** 97
Rosemarie D.L.C. Bernabe

9   **Fate of Clinical Research Studies After Ethical Approval—Follow-Up of Study Protocols Until Publication** 109
    Anette Blümle, Joerg J. Meerpohl, Martin Schumacher, and Erik von Elm

10 **Do Editorial Policies Support Ethical Research? A Thematic Text Analysis of Author Instructions in Psychiatry Journals** 125
    Daniel Strech, Courtney Metz, and Hannes Kahrass

Part III   Improving Common Domains of Research Governance

11 **Ensemble Space and the Ethics of Clinical Development** 137
    Jonathan Kimmelman and Spencer Phillips Hey

12 **Rethinking Risk–Benefit Evaluations in Biomedical Research** 153
    Annette Rid

13 **Towards an Alternative Account for Defining Acceptable Risk in Non-beneficial Paediatric Research** 163
    Sapfo Lignou

14 **Big Biobanks: Three Major Governance Challenges and Some Mini-constitutional Responses** 175
    Roger Brownsword

15 **Ethical Dimensions of Dynamic Consent in Data-Intense Biomedical Research—Paradigm Shift, or Red Herring?** 197
    Bettina Schmietow

16 **Using Patent Law to Enforce Ethical Standards: Proposal of a New Patent Requirement** 211
    Jan-Ole Reichardt

Chapter 1
Introduction

Daniel Strech and Marcel Mertz

Abstract There is no doubt that the current state of the art in medical care and public health provision can be traced back to the enormous increase in biomedical research over the past 100 years. "Biomedical research" here is used as a broad term, covering basic, translational, clinical, and post-authorization research towards increasing an understanding of causes, development, and effects of diseases and developing and improving preventive, diagnostic, and therapeutic interventions. Since we are still confronted with known and, to some extent, new diseases, there is an ongoing need for high-grade biomedical research—as declared by many stakeholders and underlined in national and international guidelines (CIOMS 2002; WMA 2013); this is despite the fact that new (technological) possibilities of conducting biomedical research and longstanding research procedures both involve legal and ethical challenges. It is therefore no surprise that the same stakeholders and guidelines that highlight the need for biomedical research also stress the need to protect research participants and enhance the ethical and scientific quality of research by means of a set of governance tools.

There is no doubt that the current state of the art in medical care and public health provision can be traced back to the enormous increase in biomedical research over the past 100 years. "Biomedical research" here is used as a broad term, covering basic, translational, clinical, and post-authorization research towards increasing an understanding of causes, development, and effects of diseases and developing and

---

D. Strech (✉)
Institute for History, Ethics and Philosophy of Medicine, Hannover Medical School, Carl-Neuberg-Str. 1, Hannover 30625, Germany
e-mail: strech.daniel@mh-hannover.de

M. Mertz
Institute for History, Ethics and Philosophy of Medicine, Hannover Medical School, Carl-Neuberg-Str. 1, Hannover 30625, Germany

Research Unit Ethics, Institute for History and Ethics of Medicine, University Hospital Cologne, Albertus-Magnus-Platz, Cologne 50923, Germany
e-mail: mertz.marcel@mh-hannover.de

© Springer International Publishing Switzerland 2016
D. Strech, M. Mertz (eds.), *Ethics and Governance of Biomedical Research*, Research Ethics Forum 4, DOI 10.1007/978-3-319-28731-7_1

improving preventive, diagnostic, and therapeutic interventions. Since we are still confronted with known and, to some extent, new diseases, there is an ongoing need for high-grade biomedical research—as declared by many stakeholders and underlined in national and international guidelines (CIOMS 2002; WMA 2013); this is despite the fact that new (technological) possibilities of conducting biomedical research and longstanding research procedures both involve legal and ethical challenges. It is therefore no surprise that the same stakeholders and guidelines that highlight the need for biomedical research also stress the need to protect research participants and enhance the ethical and scientific quality of research by means of a set of governance tools.

Governance of biomedical research can be understood as an umbrella term that covers the following: (a) the rather narrow field of research regulations in the sense of laws and legal authorities or oversight bodies; and (b) the broader field of guidelines, e.g. WMA (2013) or CIOMS (2002), advisory boards, editorial policies, ethics codes, and public involvement activities and other efforts that exist to promote the ethical conduct, social value, and appropriate freedom of biomedical research (DH 2005; EC 2012; OECD 2012). Governance activities in the field of biomedical research aim to achieve the following: (a) define and communicate requirements and standards; (b) provide mechanisms to ensure that these are understood and followed; and (c) monitor quality and assess adherence to standards (DH 2005).

Regulations and governance infrastructure are all based—even though sometimes only implicitly—on core values and normative principles. These include the following: safeguarding social value and scientific validity of research (making research worthwhile); enabling a favourable risk-benefit ratio for research participants (countering excessive maleficence); allowing for independent review of study protocols (e.g., preventing questionable conflicts of interest); ensuring informed consent and a fair selection of participants (avoiding discrimination and exploitation); maintaining respect for participants during research (promoting health and their sense of self-worth); and establishing collaborative partnerships, including the fair and transparent dissemination of the research results (Emanuel et al. 2000, 2008a, b).

But where do such values and normative principles for research originate, and where are they formulated, discussed, analysed and evaluated? From an academic perspective (that is to say, from a research perspective!), reflection upon ethical standards and the implications of research falls under the purview of the traditionally philosophical, but increasingly interdisciplinary, field of *research ethics*. Research ethics can be described as a sub-discipline of a more general "science ethics" (we use "science" in this context as a general term covering natural science, as well as social science and the humanities). Such a "science ethics" also considers ethical questions related to the use of scientific knowledge, the proper role of science in society, the core values, or ethos, of science, and other ethical aspects related to scientific institutions (e.g. teaching, or professional conduct as a supervisor). Research ethics, though, mostly confines its reflections to that part of the scientific enterprise that is arguably the most characteristic of science: the production of new and reliable knowledge by systematic, methodical means.

It is interesting to note that there has never been a "general research ethics" that could be applied to all scientific research. In fact, most research ethics are rather specialised—in our case, for clinical and translational biomedical research—even if there are common themes that may emerge in different research settings, such as the question of informed consent by human subjects.

However, as ethical challenges in research are intertwined with the kind of research done, the objects of research examined and the questions posed, this specialisation is not surprising. Also, it is not surprising that research that on the one hand can pose severe risks to (human) subjects, but on the other hand could also have great potential benefit, evokes more intense ethical reflection than research that does not have such implications. For this reason, it is no wonder that clinical and translational research ethics is a relatively mature branch of research ethics: the risk of harm and potential for benefit are straightforward when it comes to biomedical research.

Research ethics in biomedical research has a long history—a history which also illustrates the interrelationship of ethics and law, as their respective principles and rules often overlap in certain core issues. In Germany, for example, standards of conduct were produced as early as 1900, in reaction to public outcry concerning an experiment with syphilis in a hospital performed without the participants' consent ("'Ministerielle Anweisung an die Vorsteher der Kliniken, Polikliniken und sonstigen Krankenanstalten', 29.12.1900"; Minister der geistlichen, Unterrichts- und Medizinal-Angelegenheiten 1901). In 1931, a "Reichsrichtlinie" ("Guideline of the Reich"; Reichsminister des Inneren 1931) introduced standards for informed consent by participants or their legal proxies, particularly in respect of research involving children. Unfortunately, the dawn of the Third Reich lead to the suspension of these standards, as seen in the infamous examples of human experimentation at e.g. Dachau, Ravensbrück and Auschwitz for both military and civilian medical purposes. The experiences of these crimes of the Nazi regime led to a crucial code of research ethics and research law, namely the Nuremberg Code (1947; see e.g. Annas/Grodin 2008) and, subsequently, the first version of the well-known Declaration of Helsinki (1964; current version: see WMA 2013).

In a way, history has repeated itself when it comes to other important historical and current documents of research ethics and laws: the main impulse behind the landmark Belmont Report (1979; see e.g. Beauchamp 2008), for example, was the ethical scandal surrounding the so-called Tuskegee syphilis study (1932–1972; see Jones 2008), and the new pharmaceutical legislation ("Arzneimittelrecht") of 1976 in Germany was strongly informed by the Contergan (Thalidomide) affair of the 1960s. (For further information about the history of clinical research ethics, see e.g. *The Oxford Textbook of Clinical Research Ethics*, Emanuel et al. 2008a).

It seems that, at least in the past, new regulations in research ethics and research laws have more often been established retroactively than proactively. Therefore, it is all the more important to address the governance of biomedical research prior to (further) scandals that harm participants or put them at unnecessary risk, and that, in addition, harm research and science itself by reducing societal trust and hindering research. For this, it is necessary to better understand and further develop governance strategies.

Such *research on the governance* of biomedical research is conventionally conducted by using either normative/conceptual research methods (e.g., stemming from philosophy, ethics, or jurisprudence) or empirical research methods (e.g., from social or medical sciences).

The clarification and development of new domains in research governance, and moreover the justification of (new) strategies for the governance of these domains, necessitates normative/conceptual analysis and argumentation. Accordingly, conceptual research on research governance addresses, for example, theoretical challenges involved in balancing risks and benefits prior to clinical trials or challenges in consenting to biobank research by questioning whether the consent (conceptually) can ever be called "informed".

Common empirical methods used in studies of research governance include the following: (a) qualitative analysis and surveys to investigate, for example, stakeholder strategies and attitudes regarding specific normative challenges, such as the capacity of children to assent to clinical research; (b) literature and policy reviews to assess, for example, the status quo in editorial policies with regard to reporting ethics reviews and informed consent procedures; and (c) experimental tests to investigate, for example, participants' understanding of consent forms in gene transfer trials.

However, in recent decades, the value of combining normative/conceptual and empirical research methods and research findings has become clear. Thus, findings of empirical research need to be translated into normative recommendations on whether and how to modify specific governance strategies. But in order to avoid an is-ought fallacy (deducing normative recommendations from purely descriptive statements) and to arrive at normative justification, it is necessary to rely on ethical theories, values, or principles. This allows bridging the gap between empirical data and normative conclusions. In this regard, the objectives of studies on research governance have certain analogies to the objectives of translational research in general: (a) to validate promising concepts from basic research by testing under real-life conditions; and (b) to learn from such tests towards refining and further developing theoretical concepts. More specifically, studies on research governance may evaluate the performance of promising (normative) theories, principles, and governance strategies perform under real-life conditions (proof of concept) and also make it possible to identify significant barriers to implementation that must be dealt with in practice. Findings from these empirical investigations are employed to refine the concepts and governance strategies and to increase their "external validity", feasibility, and hence trustworthiness. However, findings of empirical research can also shed light on ethical challenges of which we were previously not—or insufficiently—aware. New ethical questions are inevitably defined, which in turn may stimulate new empirical hypotheses that will require testing through empirical research.

This combination of research methods is illustrated by the contributions to the present volume, which originated in an international conference on research ethics (see Preface). This book contains contributions from a group of leading scholars from multiple disciplines and countries. We decided to categorize their contributions under the following three overarching topics: (a) the introduction of new domains of research governance, e.g., discussing new ethical challenges or known challenges in a new perspective; (b) an analysis of challenges in common domains of research governance, e.g., discussing implementation barriers to or theoretical shortcomings of established normative approaches; and (c) the presentation of new strategies for improving such common domains, e.g., discussing new ways to realize established normative approaches or altogether new approaches.

The rationale of this grouping is to better highlight the aims and possible effects of research into the governance of biomedical research, rather than subsuming contributions with different aims and effects under a single heading, for example, research setting (e.g. paediatric research, biobanks etc.). Though it is always difficult in the context of an edited volume to group a broad range of contributions into topics, and arguably no grouping is ever entirely unproblematic, we think that this approach has the advantage of underlining the innovative nature of current research into research governance. Of course, this volume does not—and could not—aim to cover all newly debated governance domains, neither does it—nor could it—address all challenges in common governance domains. But it does address those special concerns that are currently the subject of controversial debate and that have not hitherto been addressed in other contributed volumes in the field (Hogle 2014; Pascuzzi et al. 2013; Schildmann et al. 2012).

Accordingly, in the part I headed "Introducing New Domains of Research Governance", three theoretical chapters investigate normative issues that are not—or only insufficiently—addressed in current research regulations and governance strategies.

First, *Govind Persad* (Stanford University, USA) considers whether *researching* cost-effective interventions is morally preferable to researching non-cost-effective interventions, e.g., because of increased social value since social value depends to some extent on cost-effectiveness. This is in contrast to more established arguments that (only) aim to prove the ethical importance of *providing* cost-effective interventions. Persad argues that we should take cost-effectiveness into account when ethically evaluating research and should discourage research on interventions that are not cost-effective—even when acknowledging challenges to this, such as the problem of enforcing cost-effectiveness norms and the difficulties of predicting either effectiveness or cost at the research stage.

Second, *Rebecca Johnson* (Princeton University, USA) explores whether, especially in phase I trials, there has been a shift from research participation of healthy persons as a form of *altruistic volunteering* to research participation as a form of *work* that is comparable to hazardous occupations, such as coal mining and firefighting. Johnson discusses what such a shift would imply for the obligations that trial sponsors bear towards trial participants. Finally, she argues that at least for some subset of participants, phase I research *is* indeed a form of work; however, according to findings of recent empirical research, it is work in which participants have little control and discretion over their daily tasks and that this is a more appropriate object of moral concern than the traditional discussion about the morality of exposing healthy participants to high risks.

Third, *Luana Colloca* (National Institutes of Health, USA) discusses state-of-the-art research in nocebo effects, which have been shown to result from negative expectations, previous experiences, and clinical encounters, as well as their clinical and ethical implications. One of those effects concerns the ethical conundrum that informing patients or participants about possible adverse consequences when participating in a randomized clinical trial may produce an undesirable harm (nocebo) — even though the right to be informed about potential risks and side effects is ethically beyond controversy. This has, as Colloca argues, a considerable effect on patient-clinician communication.

The part II "Challenges in Common Domains of Research Governance" consists of six in-depth explorations of ethical issues in known areas of research regulations and governance strategies.

First, *Sara Anna Suzan Dekking, Rieke van der Graaf, Martine C. de Vries, Marc B. Bierings*, and *Johannes J.M. van Delden* (University Medical Center Utrecht, the Netherlands) examine whether in paediatric oncology sharp distinctions can — or should — still be drawn between research and care. Dekking et al. analysed two recent Dutch protocols for children with acute lymphoblastic leukaemia (ALL) that have been differently categorized — one as research (ALL-11), the other as treatment (ALL-10). The authors conclude that in the current ethical paradigm, both protocols fall within the range of research, while also clearly exhibiting the objective of delivering best current treatment. Despite supporting an integrated model of care and research, Dekking et al. deem it too early to abandon the distinction between research and care in paediatric oncology.

Second, *Imme Petersen* and *Regine Kollek* (University of Hamburg, Germany) consider a different ethical problem in paediatric oncology, namely consent procedures for research participation. The authors deplore the lack of empirical studies on how parents, who usually provide consent on behalf of their children, assess whether their child should participate in research. Their chapter presents empirical findings from a population-based survey of parents in Germany whose child was diagnosed with cancer. Petersen and Kollek address questions relating to what parents think about the requirement of seeking assent, how to assess their child's competence in providing assent, who should be responsible for providing assent, and how to deal with a child's refusal to participate.

A normative proposal to the problem of children's assent in research is given by the third contribution to this section: *Jan Piasecki, Marcin Waligóra* (both Jagiellonian University, Poland), and *Vilius Dranseika* (Vilnius University, Poland) approach the problem from a consequentialist perspective. They argue that neither the capacity for making autonomous decisions regarding participation in research nor the understanding of the abstract concept of altruism should be the basis for implementing a uniform policy. The authors maintain that the benefits of a properly applied policy requiring assent from all capable children — while at the same time permitting a contribution from their parents—is more beneficial than policies that set a high- or low-age threshold for assent; this is because such an approach protects the children from harm and does not significantly slow down the process of scientific progress.

Fourth, *Rosemarie D.L.C. Bernabe* (University Medical Center Utrecht, the Netherlands) engages with ethical issues in post-authorization drug trials. Bernabe is especially concerned with the ethical reasons that allow the waiving of informed consent in different phase IV situations and examines the issue of consent waivers in terms of human rights. Her contribution also explores the relevance of decision theory and expected utility theory—specifically, multiattribute utility theory—when balancing risk and benefits in phase IV research. Bernabe argues that the increasing demands of expanding phase IV studies compel research ethics to keep abreast of this development.

Fifth, *Anette Blümle, Joeg J. Meerpohl, Martin Schumacher*, and *Erik von Elm* (University Medical Center Freiburg, Germany and Lausanne University Hospital, Switzerland) discuss the problem of under-reporting of clinical research, which can result in biased estimates of treatment effect or harm and may in turn lead to recommendations that are inappropriate or even dangerous. The authors present a cohort of clinical studies approved in 2000–2002 by the Research Ethics Committee of the University of Freiburg; they characterize the cohort, quantify its publication outcome, and compare protocols and publications with respect to selected aspects. They conclude that half of the clinical research conducted at a large German university medical centre remains unpublished. This means that research resources are probably wasted since health-care professionals, patients, and policy makers are unable to use the results when making their research- or care-related decisions.

Sixth, the study by *Daniel Strech, Courtney Metz*, and *Hannes Knüppel* (Hannover Medical School and Leibniz University of Hannover, Germany) assess the editorial policies of psychiatry journals regarding ethics review and informed consent: editors of such journals are encouraged by the International Committee of Medical Journal Editors (ICMJE) and Committee on Publication Ethics (COPE) to place requirements for informed consent and ethics review in their journal's instructions for authors. However, this is carried out in only half of psychiatry journals. Further, Strech et al. contend that even the ICMJE's recommendations in this regard are insufficient. They suggest that features of clinical studies that make them morally controversial, but not necessarily unethical, are analogous to methodological limitations and should thus be explicitly reported.

Finally, the part III "Improving Common Domains of Research Governance" comprises six chapters, which examine possible refinements and modifications in tried-and-tested realms of research regulations and governance strategies.

First, *Jonathan Kimmelman* and *Spencer Philips Hey* (McGill University, Canada) offer a new conceptual tool, called ensemble space, for answering questions about clinical development and clinical translation, especially in early phases of research. Examples of such questions are "At what point is it acceptable to substitute an unproven substance for standard care in a randomized trial?" and "What level of nontherapeutic risk is acceptable in trials of novel drugs?" Ensemble space is constituted by n dimensions, e.g., age, delivery, location, diagnostic score, and comorbidity intensity, which can be displayed graphically. The authors choose three dimensions as an example: dose response (x-axis), timing response (y-axis), and a measure of benefit and risk (z-axis). They conclude that the true test of their new tool's value will be shown if it is able to improve the moral economy of drug development by minimizing patient burden and maximizing yield, accuracy, and precision of evidence for further clinical development.

Second, *Annette Rid* (King's College London, UK) argues that despite their fundamental importance, surprisingly few risk-benefit evaluations answer the following question: "Under which conditions are the risks to individual participants acceptable in light of the potential social benefits of the research?" Rid maintains that it is time to de-emphasize the role of informed consent as a condition for acceptable research risk. Instead, in her view, we should strive to develop a comprehensive framework for risk-benefit evaluations that fundamentally revolves around the relation between individual risk and potential social benefit—and delineates several levels of acceptable risk—while assigning an important role to consent.

Third, *Sapfo Lignou* (University College London, UK) proposes an alternative means of assessing the ethical threshold of acceptable risk in paediatric research. Lignou first examines the "risks of daily life" standard, the "routine examinations" standard and the "charitable participation" standard in defining the ethical thresholds; she argues that none of them can provide a satisfactory, morally justified framework without inconsistencies. The proposed alternative defines the threshold according to the risk that parents are willing to expose their children in a vaccination programme in the case of an epidemic. Lignou argues that this proposal does not lead to the inconsistencies found with the other standards.

Fourth, *Roger Brownsword* (King's College London, UK) examines big biobanks. Much like a library, these population-wide biobanks are established as a resource that can be curated for access and use by the research community, and that, because of the rapidly falling cost of gene sequencing, will likely have an important place in future health-related research. Brownsword focuses on three challenges of governing such biobanks: the possibility of a functioning individual informed consent process; the question of any responsibility of biobanks or researchers working with them to return individual clinically significant findings to participants (who remain identifiable because of the longitudinal nature of such research); and how the aspiration that big biobanks be in the "public interest" is to be understood. He concludes that existing governance frameworks are unable to cope with the problems caused by existing big biobanks, but that we have the opportunity to set up better frameworks for future biobanks.

The fifth contribution, from *Bettina Schmietow* (Nuffield Council on Bioethics, UK[1]), also examines issues related to biobanks; however, she concentrates on the

---

[1] Bettina Schmietow is currently research officer at the Nuffield Council on Bioethics, UK, and previously finished the PhD programme in "Foundations and Ethics of the Life Sciences" at the European School of Molecular Medicine and in cooperation with the European Institute of Oncology and the University of Milan, Italy. Her contribution in this volume is associated with this former occupation.

question of whether informed consent can and should adapt to such emerging forms of research and whether broader or even "open" forms of consent redefine the role of autonomy and privacy—instead of proposing an accountable kind of governance. The active engagement of the general public in research governance, which relies on mass data input, and the accompanying adaptive tendency of consent in current bio-governance may, as Schmietow argues, constitute a paradigm shift towards more equal relationships between researchers and participants; however, it could also be a kind of red herring that tends to disregard potentially problematic issues relating to digitalized genomic research that is based on data sharing.

Sixth and finally, *Jan-Ole Reichardt* (University of Münster, Germany) discusses if there is a way to protect study participants in developing countries by adopting a pragmatic approach that uses the incentivizing mechanism of our patenting regimes. Because clinical trials are expensive, it is a common measure of cost-saving for many trials to be relocated to developing countries; but there, the protection of study participants is often precarious. Reichardt argues that by linking the granting of economic benefits via patents to the fulfilment of ethical requirements when conducting a trial that eventually will lead to a patent, it could be possible to compel researchers and their funding organisations to adhere to high ethical standards even when trials are relocated to developing countries.

References

Annas, G.J., and M.A. Grodin. 2008. The nuremberg code. In *The oxford textbook of clinical research ethics*, ed. E.J. Emanuel, C. Grady, R.A. Crouch, R.K. Lie, F.G. Miller, and D. Wendler, 136–140. Oxford: Oxford University Press.

Beauchamp, T.L. 2008. The Belmont report. In *The oxford textbook of clinical research ethics*, ed. E.J. Emanuel, C. Grady, R.A. Crouch, R.K. Lie, F.G. Miller, and D. Wendler, 149–155. Oxford: Oxford University Press.

Council for International Organizations of Medical Sciences (CIOMS). 2002. *International ethical guidelines for biomedical research involving human subjects*. Geneva: Council for International Organizations of Medical Sciences.

Department of Health (DH). 2005. *Research governance framework for health and social care*. https://www.gov.uk/government/uploads/system/uploads/attachment_data/file/139565/dh_4122427.pdf. Accessed 13 Jan 2015.

Emanuel, E.J., D. Wendler, and C. Grady. 2000. What makes clinical research ethical? *JAMA* 283(20): 2701–2711.

Emanuel, E.J., C. Grady, R.A. Crouch, R.K. Lie, F.G. Miller, and D. Wendler (eds.). 2008a. *The oxford textbook of clinical research ethics*. Oxford: Oxford University Press.

Emanuel, E.J., D. Wendler, and C. Grady. 2008b. An ethical framework for biomedical research. In *The oxford textbook of clinical research ethics*, ed. E.J. Emanuel, C. Grady, R.A. Crouch, R.K. Lie, F.G. Miller, and D. Wendler, 123–135. Oxford: Oxford University Press.

European Commission (EC). 2012. *Biobanks for Europe. A challenge for governance*. http://www.coe.int/t/dg3/healthbioethic/Activities/10_Biobanks/biobanks_for_Europe.pdf. Accessed 13 Jan 2015.

Hogle, L.F. (ed.). 2014. *Regenerative medicine ethics: Governing research and knowledge practices*. New York: Springer.

Jones, J.H. 2008. The Tuskegee Spyhilis experiment. In *The oxford textbook of clinical research ethics*, ed. E.J. Emanuel, C. Grady, R.A. Crouch, R.K. Lie, F.G. Miller, and D. Wendler, 86–96. Oxford: Oxford University Press.

Minister der geistlichen, Unterrichts- und Medizinal-Angelegenheiten. 1901. Anweisung an die Vorsteher der Kliniken, Polikliniken und sonstigen Krankenanstalten. In *Centralblatt für die gesamte Unterrichts-Verwaltung in Preußen 2*, ed. Ministerium der geistlichen, Unterrichts- und Medizinal-Angelegenheiten, 188–189. Berlin: Cotta.

Organisation for Economic Co-Operation and Development (OECD). 2012. *OECD recommendation on the governance of clinical trials*. http://www.irdirc.org/wp-content/uploads/2013/08/oecd-recommendation-governance-of-clinical-trials.pdf. Accessed 13 Jan 2015.

Pascuzzi, G., U. Izzo, and M. Macilotti. 2013. *Comparative issues in the governance of research biobanks: Property, privacy, intellectual property, and the role of technology*. New York: Springer.

Reichsminister des Inneren. 1931. Reichsrichtlinien zur Forschung am Menschen. *Reichsgesundheitsblatt* 6(65):174–175.

Schildmann, J., V. Sandow, O. Rauprich, and J. Vollmann. 2012. *Human medical research: Ethical, legal and socio-cultural aspects*. New York: Springer.

World Medical Association (WMA). 2013. Declaration of Helsinki: Ethical principles for medical research involving human subjects. In *64th WMA general assembly*. Fortaleza, Brazil. http://www.wma.net/en/30publications/10policies/b3. Accessed 13 Jan 2015.

Part I
Introducing New Domains of Research Governance

Chapter 2
Should Research Ethics Encourage the Production of Cost-Effective Interventions?

Govind Persad

Abstract This project considers whether and how research ethics can contribute to the provision of cost-effective medical interventions. Clinical research ethics represents an underexplored context for the promotion of cost-effectiveness. In particular, although scholars have recently argued that research on less-expensive, less-effective interventions can be ethical, there has been little or no discussion of whether ethical considerations justify curtailing research on *more expensive, more effective* interventions. Yet considering cost-effectiveness at the research stage can help ensure that scarce resources such as tissue samples or limited subject populations are employed where they do the most good; can support parallel efforts by providers and insurers to promote cost-effectiveness; and can ensure that research has social value and benefits subjects. I discuss and rebut potential objections to the consideration of cost-effectiveness in research, including the difficulty of predicting effectiveness and cost at the research stage, concerns about limitations in cost-effectiveness analysis, and worries about overly limiting researchers' freedom. I then consider the advantages and disadvantages of having certain participants in the research enterprise, including IRBs, advisory committees, sponsors, investigators, and subjects, consider cost-effectiveness. The project concludes by qualifiedly endorsing the consideration of cost-effectiveness at the research stage. While incorporating cost-effectiveness considerations into the ethical evaluation of human subjects research will not on its own ensure that the health care system realizes cost-effectiveness goals, doing so nonetheless represents an important part of a broader effort to control rising medical costs.

G. Persad (✉)
Department of Health Policy and Management, Johns Hopkins University,
Baltimore, MD 21205, USA
e-mail: gpersad@alumni.stanford.edu

2.1 Introduction

The moral importance of cost-effectiveness has gained prominence in recent debates about funding medical care. Toby Ord at Oxford, for instance, has argued that there is a moral imperative to use public funding to provide the most cost-effective interventions, such as preventive care for blindness-causing infections like trachoma, rather than more expensive and less cost-effective interventions such as guide dogs (Ord 2013). Recent debates over the cost-effectiveness of programs like the President's Emergency Plan for AIDS Relief (PEPFAR) raise similar issues (Denny and Emanuel 2008; Emanuel 2012a).

This project considers whether *researching* cost-effective interventions is morally desirable, in contrast to the above arguments, which aim to establish the ethical importance of *providing* cost-effective interventions. The view that research must have social value in order to be ethical has been prominently defended (Emanuel et al. 2000). If social value depends to some extent on cost-effectiveness, as many believe, then the cost-effectiveness of the intervention being studied will affect whether research into that intervention is ethical. Research will then serve as an institutional gatekeeper in the service of cost-effectiveness, just as many have argued that governments, insurers, physicians, and patients should work to promote interventions that are cost-effective over those that are not.

I argue for a stance of cautious optimism toward proposals that we consider cost-effectiveness when evaluating the ethics of proposed research interventions. But I believe that using research ethics to focus research on interventions that are cost-effective faces several challenges, most importantly that (1) enforcing cost-effectiveness norms and (2) predicting either effectiveness or cost are particularly difficult at the research stage. Notwithstanding these challenges, research ethics constitute an underexplored and potentially important part of the enterprise of promoting cost-effectiveness in medicine.

2.2 Cost-Effectiveness in the Development of New Interventions

Cost-effectiveness analysis in medicine involves comparing the *cost* of medical interventions, such as pharmaceuticals or devices, against the *effectiveness* of these interventions at producing a desired health outcome. Policymakers and medical ethicists have argued that medicine should place a greater emphasis on promoting the use of interventions that are cost-effective and discouraging the use of those that are not (Emanuel and Fuchs 2008; Orszag and Ellis 2007; Mortimer and Peacock 2012). In circumstances of scarcity, some argue that cost-effectiveness rises to the level of a moral imperative: spending a limited pool of money on interventions that are cost-effective enables us to meet more health needs (Denny and Emanuel 2008; Ord 2013).

The provision of medical interventions to patients represents the culmination of a multi-stage process involving many actors, and providing many avenues for policy initiatives to promote cost-effectiveness. Consider, for example, the development of lovastatin, the first of the widely prescribed cholesterol-lowering "statin" drugs. During the 1970s, basic science research uncovered the biosynthetic pathways by which humans synthesize cholesterol and discovered that compounds such as lovastatin inhibit that pathway in fungal models and in preclinical research on animals; during the 1980s, lovastatin entered clinical trials; and in 1987, lovastatin was approved by the FDA in the United States (Tobert 2003). Lovastatin was quickly accepted by physicians and patients, although some hospitals and insurers attempted to restrict its use as a first-line treatment due to its high cost at the time (Grabowski 1998; Lederle and Rogers 1990). The story of lovastatin illustrates the pathway from the discovery of a promising compound to the provision of a medical intervention. Although cost-effectiveness considerations only entered the process at the hospital or insurer stage in the case of lovastatin, cost-effectiveness considerations can and do enter at upstream and downstream stages as well (Table 2.1). Downstream, value-based insurance designs could encourage patients to choose cost-effective interventions (Thomson et al. 2013). Codes of ethics could encourage physicians to provide cost-effective interventions (Emanuel 2012b; Weinberger 2011). Upstream, governmental bodies could provide cost-effectiveness information to insurers, hospitals, and physicians (Wilensky 2006). Regulatory agencies could consider an intervention's cost-effectiveness when deciding whether to approve it for marketing

Table 2.1 Promoting cost-effectiveness in the development of new interventions

Stage	Actors	Strategies to promote cost-effectiveness
Basic science research	Scientists	Prioritize funding for cost-effective interventions
	Sponsors	
Preclinical and clinical trials	Investigators	Prioritize funding for cost-effective interventions
	Sponsors	Consider cost-effectiveness when approving trials
	Research subjects	Choose not to pursue research on cost-ineffective interventions
	Research ethics committees	Refuse to participate in trials of interventions that are not cost-effective
Approval for marketing	Regulatory agencies (e.g. FDA)	Consider cost-effectiveness in approval process
Approval for reimbursement	Insurers (e.g. Blue Cross, Medicare)	Reimburse based on cost-effectiveness
		Tax insurance that provides cost-ineffective interventions
Use in clinical practice	Hospitals, physicians	Use formularies
Use with specific patients	Physicians, patients	Educate physicians about cost-effective practice
		Adopt value-based insurance that incentivizes patients to choose cost-effective interventions

(Paltiel and Pollack 2010). And—as this chapter discusses—investigators, sponsors, research ethics committees, and research participants could consider an intervention's cost-effectiveness when deciding whether to begin a clinical trial.

2.3 Why Promote Cost-Effective Interventions at the Research Stage?

Cost-effectiveness considerations are not altogether foreign to debates over which interventions should advance to human testing. However, the most active debates about the relevance of cost-effectiveness to human-subject research focus on whether cost-effectiveness considerations can *expand*, rather than *limit*, the scope of allowable research. These debates arise in response to proposed research on "less expensive, less effective" interventions: those that promise to be more cost-effective than the *status quo*, at some sacrifice to absolute effectiveness. Examples include research on less costly methods of lead abatement (Buchanan and Miller 2006), less costly treatment for multiple sclerosis (Lie 2004), replacement of multidrug regimens by monotherapy for HIV (Girardi and Angeletti 2013), and less costly prevention of maternal-fetal HIV transmission (Wendler et al. 2004).

That an intervention's cost-effectiveness can make research that would otherwise be disallowed ethical, as many of the above authors argue, suggests that cost-effectiveness has moral significance in the research context. This paper explores the heretofore ignored "flip side" of the debate above: whether an intervention's *lack* of cost-effectiveness can render research into that intervention ethically objectionable.

Dividing interventions into the four quadrants of the cost-effectiveness plane (Black 1990) helps to illustrate the relationship between the "less expensive, less effective" debate and the questions explored in this paper (Table 2.2).

Research on "dominant" interventions, those promising to be both more cost-effective and more effective than the *status quo*, seems clearly acceptable. Research on "dominated" interventions, those less cost-effective and less absolutely effective than the *status quo*, seems clearly unacceptable. The question I explore here is whether we can treat research in the northeast quadrant as unproblematic while strenuously debating research in the southwest quadrant. Other commentators have argued for more parallel treatment of the two quadrants at the stage of reimbursement and prescribing decisions (Dowie 2004) but this debate has not so far extended to research.

Table 2.2 Research ethics and the cost-effectiveness plane

	More cost-effective	Less cost-effective
More absolutely effective	Consensus permission to research	Consensus permission to research, challenged here
Less absolutely effective	Research permissibility debated	Consensus prohibition on research

The numerous policy proposals discussed in Part I—from physician education to value-based insurance to cost-effectiveness thresholds for reimbursement—all restrict the *provision* of interventions that are not cost-effective, even though they may be absolutely effective. In contrast, the current consensus in research ethics seems to treat *research* into interventions that are more absolutely effective but less cost-effective as ethically unproblematic. Yet considering cost-effectiveness in research could improve the use of scarce resources, reduce pressure on actors at downstream stages, and ensure that research risks to subjects are appropriately counterbalanced by social benefits.

2.3.1 Appropriate Use of Scarce Resources

First, clinical research—like the provision of interventions—occurs within a broader context of scarcity. Some research requires the use of scarce medical resources such as fetal tissue (Woods and Taylor 2008). Other research, such as research on pediatric mood disorders, can only be performed on a limited population of subjects, which makes subjects a scarce resource (Frank et al. 2002). Some have argued that research subjects are a scarce resource in general (London et al. 2013; Dresser 2012). Where scarcity exists, priorities must be set, and cost-effectiveness considerations can help us use limited research resources to produce interventions that will help more patients.

2.3.2 Supporting Downstream Actors

Second, it may be easier to restrict the provision of treatment on cost-effectiveness grounds at earlier stages, such as research or approval, than at later stages, such as the physician–patient interaction (Garber 1994). First, those who have invested time and money in researching the intervention will have a stake in lobbying against cost-effectiveness restrictions, as pharmaceutical manufacturers did after the Australian national health insurance agency declined to cover a cervical cancer vaccine (Roughead et al. 2008). Second, the further the intervention progresses, the more likely it becomes that physicians and patients will come to expect or aspire to receive the new intervention; if they treat these expectations or aspirations as their new baseline, they will frame the denial of reimbursement as a psychologically more upsetting loss rather than a mere failure to gain. For instance, it is difficult to cease provision of a drug that was being provisionally provided while its effectiveness was assessed, even if the drug proves ineffective. (While restrictions on basic science also have these attractions, the multi-purpose nature of basic science means that it will be harder to target restrictions without interfering with research that may yield cost-effective interventions.)

2.3.3 Social Value and Fairness to Research Subjects

Finally, research involves exposing subjects to risk and harm, which requires countervailing benefits. Even at the stage of preclinical research on animals, ethical guidelines require that the research have social value (Prentice et al. 1992); likewise, Emanuel, Grady, and Wendler propose that research on human subjects must have social value to be ethical (Emanuel et al. 2000). Some propose that research must not only have social value but also be "responsive" or provide "reasonable availability"—that subjects must have a reasonable prospect of benefiting from the intervention being researched (London 2008).

Cost-effectiveness can help contribute to social value, responsiveness, and reasonable availability. For instance, the expected cost-effectiveness of an HPV vaccine regimen can contribute to its social value at the research stage (Lindsey et al. 2013), and the cost-effectiveness of a hemophilia treatment can establish its reasonable availability to participants (Dimichele 2008). Costly research on gene therapy with little evidence of benefit may lack social value (King 2003). Indeed, if we accept a reasonable availability or responsiveness requirement, then human subjects research on interventions that subject populations will never receive because later-stage gatekeepers (such as physicians, approval bodies, or insurers) will not provide them on cost-effectiveness grounds is unethical.

Despite the above arguments, there have been few proposals to incorporate cost-effectiveness standards into research ethics. One exception is Christine Grady and Tito Fojo's recent criticism of cancer treatments that provide small benefits at very high cost, which included a proposal that research be limited on cost-effectiveness grounds (Fojo and Grady 2009):

> Research studies that are powered to detect a survival advantage of 2 months or less should only test interventions that can be marketed at a cost of less than $20 000 for a course of treatment, which is a monetary value consistent with the cost of one quality adjusted life year in patients treated with artificial renal dialysis ($129 090). Similarly, a study designed to detect a 4-month advantage can test a therapy that will cost up to $30 000 per patient.

Although Grady and Fojo's proposal generated a great deal of discussion, very little of the discussion focused specifically on their proposed limitation on research. Yet the arguments discussed above give several reasons in support of such a limitation. In the next section, I consider some objections to the consideration of cost-effectiveness in research.

2.4 Objections to Considering Cost-Effectiveness at the Research Stage

2.4.1 The Unpredictability of Effectiveness

One major objection to the use of cost-effectiveness considerations in research involves the difficulty of predicting, at the clinical trial stage, the effectiveness of the intervention being researched. Assessing cost-effectiveness requires an accurate measure of effectiveness. Two critics of Grady and Fojo's proposal identified this point (Cohen and Looney 2010):

> Determining the drug's clinical value is not something that can or should be decided before a drug's approval, in part because this is what markets do after approval but also because of the considerable uncertainty associated with a drug's real-world effectiveness.

However, even if effectiveness cannot be as precisely predicted at the research stage as at later stages, there are ways of assessing effectiveness at the trial stage. First, as Grady and Fojo point out, we can use the research study's power as an outer bound on effectiveness: if a study is powered to detect a 2-month survival difference, then the study cannot show the intervention to be any more effective than achieving a 2-month increment in survival. Second, trial designs have been devised that combine cost-effectiveness predictions with determinations of clinical efficacy (Briggs 2000; Drummond and Stoddart 1984).

2.4.2 The Unpredictability of Cost

The other dimension of cost-effectiveness measures—cost—is also challenging to predict at the clinical trial stage. Cohen and Looney likewise therefore argue that "[i]t is inappropriate for pharmaceutical sponsors to impose de facto price controls on themselves before a drug's approval" (Cohen and Looney 2010). While an intervention's clinical effectiveness is largely determined by human biology, its cost is limited only by human choice. The interventions Grady and Fojo claim we should not research, like cetuximab, could become highly cost-effective if their price was lowered dramatically.

The measurement of drug costs in cost-effectiveness analysis is controversial. The economic ideal is to compare the opportunity cost of the intervention to its benefits (Garrison et al. 2010). However, the market price of a medical intervention does not invariably reflect its opportunity cost. For instance, some payments to providers reflect the effect of patent rents (Palmer and Raftery 1999) or market distortions (Neumann 2009). Estimating costs is particularly difficult at the research stage because so little information is available.

Notwithstanding these difficulties, some cases exist where opportunity costs are easily identifiable even at the research stage, because at least one resource in question is absolutely and immediately scarce. Multiple-organ transplants, for instance, arguably constitute a cost-ineffective use of a scarce resource: they use up multiple organs to save one life when those organs could have saved two or three (Menzel 1994). As such, if we accept the argument against *providing* multiple-organ transplants, we should also accept a parallel argument against *researching* such transplants. Likewise, some argued that pharmaceutical companies acted inappropriately in promoting the use of antibiotics to treat less severe conditions when such use would produce resistance that jeopardizes public health (Kesselheim and Outterson 2010). If this is true, research on the efficacy of antibiotics in treating less severe conditions would also be unethical.

The more difficult questions involve whether and how we should evaluate the costs of an intervention at the research stage where absolute scarcity does not exist. This involves predicting the various costs of the intervention, such as the time it will take health professionals to administer it, the wages and profits that will need to be paid to those who develop and provide it, and the cost of raw materials and technical equipment that will be used in developing and administering it. While some of these predictions are technically challenging, empirical and conceptual work on priority setting in research shows promise in helping predict and weigh many of the above costs (Bojke et al. 2007; Fleurence and Torgerson 2004; Torgerson 2002; Rudan 2012).

An important issue in prediction is that some costs—in particular wages and profits—arise from discretionary choices. Pharmaceutical companies could choose to accept lower profits than they do (Schüklenk 2002) or physicians lower incomes (Curzer 1992; Menzel 1983). In particular—as Grady and Fojo suggest—trial sponsors could be required to accept a "cost ceiling" that guarantees that a proposed intervention will be made available at a specified price before research is allowed to proceed. Such a cost ceiling would be analogous to post-trial access requirements that have been imposed in developing-country trials (Grady 2005). If manufacturers refuse to accept a cost ceiling, this undermines their complaint that their research is being unjustly limited on cost-effectiveness grounds, since their demands are what prevent the interventions from being cost-effective. Manufacturers who demand high prices and then complain about limits would be analogous to kidnappers who complain about the hard-heartedness of government refusals to ransom kidnapped individuals (G. Cohen 2010).

Some might worry that forcing manufacturers to direct research toward cost-effective interventions will lead to insufficient research into useful medical interventions. However, current trends suggest an excess of research into costly interventions with marginal benefit and a deficit of research on cost-effective interventions (Yamey 2002), so encouraging manufacturers to align their research efforts with cost-effective goals seems warranted.

2.4.3 Equity Concerns About Cost-Effectiveness Analysis

Some have worried that cost-effectiveness analysis is insensitive to concerns about justice and interpersonal equity (Brock 2004); others have gone so far as to entirely reject its use on those grounds (Harris 2005). While cost-effectiveness analysis has flaws, they cannot justify adopting an approach to research, or to any other stage of the process, that entirely ignores cost-effectiveness considerations (Mortimer and Peacock 2012). Rather, cost-effectiveness analysis can and should be improved to take equity values into account (Menzel 1999).

Likewise, some argue that before imposing limits on interventions that are not cost-effective, we must first address wasteful spending elsewhere, such as in national defense (Angell 1985; Daniels 1986; Schrecker 2013). However, while wasteful defense spending may *mitigate* the culpability of medical researchers who pursue research on cost-ineffective interventions for the resulting avoidable morbidity and mortality, it does not *eliminate* their culpability, just as criminals' culpability does not eliminate crime victims' responsibility to avoid injuring innocent bystanders in self-defense (Hurka 2005; McMahan 2011). Indeed, as Harry Frankfurt has argued, two actors can both be fully responsible for an outcome even when neither of their acts alone would have been sufficient to produce the outcome (Frankfurt 1971).

2.4.4 Freedom of Intellectual Inquiry

A final objection is that research is a form of intellectual inquiry, and that requiring researchers to consider cost-effectiveness will stifle free inquiry. While this objection has force in the context of basic science research, it has much less force where human-subject research is concerned. Conducting human-subject research is a privilege that comes with conditions and outside oversight, not a purely private matter between investigator and subject (Dresser 2012). The use of human subjects in research—even with informed consent—requires that the research have social value (Emanuel et al. 2000; Joffe and Miller 2008).

2.5 Who Should Ensure That Research Promotes the Development of Cost-Effective Interventions?

If we accept the ethical legitimacy of incorporating cost-effectiveness considerations at the research stage, we face the challenging question of how cost-effectiveness limitations on research should be implemented. Grady and Fojo, for example, do not say who should decide not to pursue clinical trials on interventions that are not cost-effective.

2.5.1 Research Ethics Committees

Research ethics committees, such as institutional review boards (IRBs), could employ cost-effectiveness judgments as part of their evaluation of whether the research in question will have social value. For instance, Berg et al. suggest that it might be appropriate to deny approval for research into enhancement technologies if such technologies prove to lack cost-effectiveness (Berg et al. 2009). Because research ethics committees already have enforcement power and technical expertise, and already review for scientific validity, they seem a natural gatekeeper for ensuring that research has social value (London et al. 2013). However, assessing social value arguably lies outside the core expertise of ethics review committees (Rid and Wendler 2010), and many research ethics committees are already overloaded.

Some may argue that an IRB that refuses to approve research on the basis that the interventions produced will not be cost-effective engages in an assessment of the "possible long-range effects of applying knowledge gained in the research," which United States law bars IRBs from engaging in (Mano et al. 2006). However, the legislative intent of this provision was to prevent IRBs from stifling research on controversial topics, such as correlations between race or gender and cognitive ability or criminality (London et al. 2013; Mehlman and Berg 2008). Assessments of cost-effectiveness focus on the *importance* of the knowledge the research will provide, which IRBs are permitted to assess, rather than the *social popularity* of the research, which IRBs must not consider.

Finally, the alignment between libertarian objections to IRB review and libertarian objections to regulation of health care by organizations like the FDA (Epstein 2007) suggests that the same political backlash that has prevented the use of cost-effectiveness analysis at other stages (Neumann and Weinstein 2010) may also hamper IRBs in integrating cost-effectiveness norms into research ethics. IRBs attempting to limit human subjects research on cost-effectiveness grounds may be accused of "mission creep" and censorship, as they have been in other contexts (Hyman 2007). IRBs' remarkable insulation from political intervention (Zywicki 2007), however, suggests that they may well succeed where other, more elegant institutional homes for the promulgation of cost-effectiveness norms have not.

2.5.2 Research Advisory Committees

Given IRBs' local focus and their limited expertise and legal authority, several authors have proposed that research advisory committees (RACs) or other central bodies investigate the social implications of research in specific areas, such as behavioral genetics, harm reduction, human enhancement, stem cell research, and post-trial access to interventions (Baylis and Scott Robert 2006; Fleischman et al. 2011; Mano et al. 2006). Research on recombinant DNA technologies is currently overseen by an RAC (King 2002).

Would an RAC be an appropriate body to integrate concerns about the cost-effectiveness of interventions into research ethics? Nancy King recommends that RACs be used where "overarching umbrella review and field-wide guidance is needed and useful; cross-study analysis of research data for a field is both possible and desirable; and public access and education are desired" (King 2002). Several of King's considerations—in particular, the need for field-wide guidance and cross-study analysis—do seem applicable to cost-effectiveness. However, cost-effectiveness considerations are relevant to *every* clinical trial, while existing and proposed RACs focus on a particular area of research, such as human enhancement or gene transfer. As such, requiring all proposals to pass through RAC review might be unworkably broad; however, a RAC playing a more advisory role might be able to collect valuable data on the cost-effectiveness of interventions under research that might in turn inform other actors involved in research decision making. A RAC with enforcement power might also be appropriate for areas of research where cost-effectiveness is uniquely problematic, as Grady and Fojo suggest is true in certain sectors of cancer research.

2.5.3 Sponsors

Another possibility is that trial sponsors should employ cost-effectiveness judgments when deciding whether or not to fund a given clinical trial. To the extent that trials are privately sponsored, sponsors' interest in producing profitable drugs may seem to militate against this option. However, if cost-effectiveness considerations are relevant at downstream stages, such as formulary inclusion, manufacturers may have an economic interest in ensuring that the interventions they research are likely to be adopted in the regulated marketplace. Furthermore, sponsors conducting human subjects research may have ethical obligations other than the maximization of profit, which may include obligations to ensure that the research they sponsor is socially valuable (Shah 2013; Spinello 1992).

2.5.4 Investigators

A fourth possibility is that investigators' codes of ethics should prohibit work on trials that produce interventions that are not cost effective. The possibility of establishing codes of ethics for investigators or revisiting existing codes of ethics has occasionally been discussed as an alternative to IRB review, or as a complement to such review (Shah 2013). A recent boycott of Abbott Laboratories' clinical trials motivated by the high costs of Abbott's HIV drugs seems to reflect some physicians' ethical concern that clinical trials are not leading to the production of cost-effective interventions (Dixon and Richwine 2004).

However, as Shah notes, investigators' ethical obligations—like sponsors' obligations—are under-discussed in the literature (Shah 2013). An investigators' code of ethics that alerts them to the ethical importance of cost-effectiveness, as some codes of ethics do for physicians (Emanuel 2012b), could help ensure that research promotes the development of cost-effective interventions. For instance, Franklin Miller and Steven Joffe's proposed code of ethics for researchers, which regards "promoting *socially valuable* knowledge about health, disease, and treatment" (emphasis added) as the central aim of human-subject research (Joffe and Miller 2008), justifies researchers' attention to cost-effectiveness to the extent that interventions that are not cost-effective are not socially valuable. Likewise, some scientific codes of ethics emphasize that scientific research must "enhance the public interest or well-being" or otherwise serve the public interest (Resnik and Shamoo 2005).

2.5.5 Research Subjects

Finally, prospective subjects could refuse to participate in trials that will not produce cost-effective interventions. For instance, breast cancer advocates have advised prospective subjects to "boycott clinical trials by companies that won't agree to price controls, and which maintain secrecy about their true R&D costs" (Batt 2000). Likewise, a recent initiative in the United Kingdom has exhorted research subjects not to participate in clinical trials that fail to guarantee that their results will be made public (Kmietowicz 2013; Limb 2013). Such initiatives might be made easier by proposals that trial participants be told about the likely future costs of the intervention under study (Barnbaum 2011).

Efforts by research subjects to promote cost-effectiveness face many of the same problems that generally plague workers and consumers attempting to organize (Lynch 2014). For instance, manufacturers can simply pay subjects enough to overcome their objections. Boycotts may end up obstructing research into interventions that are in fact cost-effective, because subjects are unlikely to have access to the most detailed information about future costs or effectiveness. And existing boycott proposals, though congruent with cost-effectiveness, do not take cost-effectiveness as their chief object. Nonetheless, especially if other actors at the research stage are unable or unwilling to assist in implementing cost-effectiveness norms, advocacy and collective action by subjects could play an important role in discouraging the production of interventions that are not cost-effective.

2.6 Conclusion

This article has presented and evaluated the arguments for and against using research ethics to encourage the production of cost-effective interventions. At this point, I can offer a tempered endorsement of doing so. Promoting the production of cost-effective interventions at the research stage will not, on its own, achieve the long-sought goal of cost control in medicine. Nor would it be wise to scale back cost control efforts at other stages and use research restrictions as the main gatekeeper. The research enterprise is not well suited to be the primary evaluator of cost-effectiveness or primary enforcer of cost-effectiveness norms.

Nonetheless, research can help to share the burden of making cost-effectiveness judgments rather than leaving such judgments to downstream actors such as the FDA, insurers, hospitals, physicians, or patients. Others have argued that physicians must be among the actors empowered to consider cost-effectiveness, because relying solely on approval bodies to contain costs will be undermined by physicians' lack of commitment to cost-effectiveness (Ubel and Arnold 1995). Similarly, if researchers are not committed to cost-effectiveness, they may generate a flood of cost-ineffective interventions that overwhelms downstream actors (Rettig 1994). In contrast, judicious adoption of cost-effectiveness norms at the research stage enlists research as part of an "all hands on deck" approach that empowers actors at every level of the scientific and regulatory process to promote the use of cost-effective interventions (Emanuel and Steinmetz 2013). Existing attempts to limit health care costs have not succeeded in stemming their rise. Promoting cost-effective interventions at the research stage could represent an important part of an experimental, multi-level approach—like that adopted in the United States under the Affordable Care Act (Orszag and Emanuel 2010)—to reining in the expanding cost of medical care.

Acknowledgment I am grateful to John Phillips for his helpful written comments on an earlier draft, and to attendees at the Conference on Current Challenges in Preclinical, Clinical, and Public Health Ethics, Hannover Medical School, 2013 for their comments.

References

Angell, M. 1985. Cost containment and the physician. *JAMA* 254(9): 1203–1207.

Barnbaum, D. 2011. You get what someone else will pay for. *Theory and Application Ethics* 1(2): 28–31.

Batt, S. 2000. The new genetic therapies: The case of Herceptin for breast cancer. In *The gender of genetic futures: the Canadian biotechnology strategy, women and health*, ed. F. Miller, L. Weir, R. Mykitiuk, P. Lee, S. Sherwin, and S. Tudiver, 9–17. Toronto: National Network on Environments and Women's Health Working Paper Series.

Baylis, F., and J.S. Robert. 2006. Human embryonic stem cell research: An argument for national research review. *Accountability in Research* 13(3): 207–224.

Berg, J.W., M.J. Mehlman, D.B. Rubin, and E. Kodish. 2009. Making all the children above average: Ethical and regulatory concerns for pediatricians in pediatric enhancement research. *Clinical Pediatrics* 48(5): 472–480.

Black, W.C. 1990. The CE plane: A graphic representation of cost-effectiveness. *Medical Decision Making* 10(3): 212–214.

Bojke, L., K. Claxton, M.J. Sculpher, and S. Palmer. 2007. Identifying research priorities: The value of information associated with repeat screening for age-related macular degeneration. *Medical Decision Making* 28(1): 33–43.

Briggs, A. 2000. Economic evaluation and clinical trials: Size matters. *BMJ* 321(7273): 1362–1363.

Brock D.W. 2004. Ethical issues in the use of cost effectiveness analysis for the prioritization of health care resources. In Making choices in health: WHO guide to cost-effectiveness analysis, ed. T. Tan-Torres Edejer, R. Baltussen, T. Adam, R. Hutubessy, A. Acharya, D.B. Evans, et al, 289–312. Geneva: World Health Organization.

Buchanan, D.R., and F.G. Miller. 2006. Justice and fairness in the Kennedy Krieger institute lead paint study: The ethics of public health research on less expensive, less effective interventions. *American Journal of Public Health* 96(5): 781–787.

Cohen, G.A. 2010. *Rescuing justice and equality*. Cambridge, MA: Harvard University Press.

Cohen, J., and W. Looney. 2010. Re: How much is life worth: Cetuximab, non-small cell lung cancer, and the $440 billion question. *JNCI* 102(15): 1207–7.

Curzer, H.J. 1992. Do physicians make too much money? *Theoretical Medicine* 13(1): 45–65.

Daniels, N. 1986. Why saying no to patients in the United States is so hard. *New England Journal of Medicine* 314(21): 1380–1383.

Denny, C.C., and E.J. Emanuel. 2008. US health aid beyond PEPFAR: The mother & child campaign. *JAMA* 300(17): 2048.

Dimichele, D.M. 2008. Ethical considerations in clinical investigation: Exploring relevance in haemophilia research. *Haemophilia* 14(s3): 122–129.

Dixon, K., and L. Richwine. 2004. *Doctors call for Abbott boycott on AIDS price hike*. London, England: Reuters. February 10.

Dowie, J. 2004. Why cost-effectiveness should trump (clinical) effectiveness: The ethical economics of the South West quadrant. *Health Economics* 13(5): 453–459.

Dresser, R. 2012. Alive and well: The research imperative. *The Journal of Law, Medicine & Ethics* 40(4): 915–921.

Drummond, M.F., and G.L. Stoddart. 1984. Economic analysis and clinical trials. *Controlled Clinical Trials* 5(2): 115–128.

Emanuel, E.J. 2012a. PEPFAR and maximizing the effects of global health assistance. *JAMA* 307(19): 2097–2100.

Emanuel, E.J. 2012b. Review of the American college of physicians ethics manual. *Annals of Internal Medicine* 156(1): 56–57.

Emanuel, E.J., and V.R. Fuchs. 2008. The perfect storm of overutilization. *JAMA* 299(23): 2789–2791.

Emanuel, E.J., and A. Steinmetz. 2013. Will physicians lead on controlling health care costs? *JAMA* 310(4): 374–375.

Emanuel, E.J., D. Wendler, and C. Grady. 2000. What makes clinical research ethical? *JAMA* 283(20): 2701–2711.

Epstein, R.A. 2007. The erosion of individual autonomy in medical decisionmaking: Of the FDA and IRBs. *The Georgetown Law Journal* 96: 559–582.

Fleischman, A., C. Levine, L. Eckenwiler, C. Grady, D.E. Hammerschmidt, and J. Sugarman. 2011. Dealing with the long-term social implications of research. *American Journal of Bioethics* 11(5): 5–9.

Fleurence, R.L., and D.J. Torgerson. 2004. Setting priorities for research. *Health Policy* 69(1): 1–10.

Fojo, T., and C. Grady. 2009. How much is life worth: Cetuximab, non-small cell lung cancer, and the $440 billion question. *JNCI* 101(15): 1044–1048.

Frank, E., A.J. Rush, M. Blehar, S. Essock, W. Hargreaves, M. Hogan, R. Jarrett, R.L. Johnson, W.J. Katon, and P. Lavori. 2002. Skating to where the puck is going to be: A plan for clinical trials and translation research in mood disorders. *Biological Psychiatry* 52(6): 631–654.

Frankfurt, H.G. 1971. Freedom of the will and the concept of a person. *Journal of Philosophy* 68(1): 5–20.

Garber, A.M. 1994. Can technology assessment control health spending? *Health Affairs* 13(3): 115–126.

Garrison Jr., L.P., E.C. Mansley, T.A. Abbott 3rd, B.W. Bresnahan, J.W. Hay, and J. Smeeding. 2010. Good research practices for measuring drug costs in cost-effectiveness analyses: A societal perspective: The ISPOR drug cost task force report—part II. *Value in Health* 13(1): 8–13.

Girardi, E., and C. Angeletti. 2013. Much cheaper, almost as good treatment: A possible approach to guarantee sustainability of HIV care? *HAART, HIV correlated pathologies and other infections*. 2013(18):175–178.

Grabowski, H. 1998. The role of cost-effectiveness analysis in managed-care decisions. *PharmacoEconomics* 14(Suppl(1)): 15–24.

Grady, C. 2005. The challenge of assuring continued post-trial access to beneficial treatment. *Yale Journal of Health Policy, Law, and Ethics* 5(1): 425–435.

Harris, J. 2005. It's not NICE to discriminate. *Journal of Medical Ethics* 31(7): 373–375.

Hurka, T. 2005. Proportionality in the morality of war. *Philosophy Public Affairs* 33(1): 34–66.

Hyman, D.A. 2007. Institutional review boards: Is this the least worst we can do? *Northwestern University Law Review* 101(2): 749–774.

Joffe, S., and F.G. Miller. 2008. Bench to bedside: Mapping the moral terrain of clinical research. *The Hastings Center Report* 38(2): 30–42.

Kesselheim, A.S., and K. Outterson. 2010. Fighting antibiotic resistance: Marrying new financial incentives to meeting public health goals. *Health Affairs* 29(9): 1689–1696.

King, N.M.P. 2002. RAC oversight of gene transfer research: A model worth extending? *The Journal of Law, Medicine & Ethics* 30(3): 381–389.

King, N.M.P. 2003. Accident & desire. Inadvertent germline effects in clinical research. *The Hastings Center Report* 33(2): 23–30.

Kmietowicz, Z. 2013. Patients are urged to boycott trials that do not guarantee publication. *BMJ* 346: f106.

Lederle, F.A., and E.M. Rogers. 1990. Lowering the cost of lowering the cholesterol: A formulary policy for lovastatin. *Journal of General Internal Medicine* 5(6): 459–463.

Lie, R.K. 2004. Research ethics and evidence based medicine. *Journal of Medical Ethics* 30(2): 122–125.

Limb, M. 2013. NICE joins campaign for full disclosure of clinical trial data. *BMJ* 346: f1269.

Lindsey, J.C., S.K. Shah, G.K. Siberry, P. Jean-Philippe, and M.J. Levin. 2013. Ethical tradeoffs in trial design: Case study of an HPV vaccine trial in HIV-infected adolescent girls in lower income settings. *Developing World Bioethics* 13(2): 95–104.

London, A.J. 2008. Responsiveness to host community health needs. In *The oxford textbook of clinical research ethics*, ed. E.J. Emanuel, R. Crouch, C. Grady, R. Lie, F. Miller, and D. Wendler, 737–746. New York: Oxford University Press.

London, A.J., J. Kimmelman, and M.E. Emborg. 2013. Beyond access vs. Protection in trials of innovative therapies. *Science* 328(5980): 829–830.

Lynch, H.F. 2014. Protecting human research subjects as human research workers. In *Human subjects research regulation*, ed. I.G. Cohen and H.F. Lynch, 327–340. Cambridge: MIT Press.

Mano, M.S., D.D. Rosa, and L.D. Lago. 2006. Multinational clinical trials in oncology and post-trial benefits for host countries: Where do we stand? *European Journal of Cancer* 42(16): 2675–2677.

McMahan, J. 2011. Proportionality in the Afghanistan war. *Ethics International Affairs* 25(2): 143–154.

Mehlman, M.J., and J.W. Berg. 2008. Human subjects protections in biomedical enhancement research: Assessing risk and benefit and obtaining informed consent. *The Journal of Law, Medicine & Ethics* 36(3): 546–549.

Menzel, P.T. 1983. *Medical costs, moral choices: A philosophy of health care economics in America*. New Haven: Yale University Press.

Menzel, P.T. 1994. Rescuing lives: Can't we count? *Hastings Center Report* 24(1): 22–23.

Menzel, P.T., M.R. Gold, E. Nord, J.L. Pinto-Prades, J. Richardson, and P. Ubel. 1999. Toward a broader view of values in cost-effectiveness analysis of health. *Hastings Center Report* 29(3): 7–15.

Mortimer, D., and S. Peacock. 2012. Social welfare and the affordable care act: Is it ever optimal to set aside comparative cost? *Social Science and Medicine* 75(7): 1156–1162.

Neumann, P.J. 2009. Costing and perspective in published cost-effectiveness analysis. *Medical Care* 47(7 Suppl 1): S28–S32.

Neumann, P.J., and M.C. Weinstein. 2010. Legislating against use of cost-effectiveness information. *The New England Journal of Medicine* 363(16): 1495–1497.

Ord, T. 2013. *The moral imperative towards cost-effectiveness in global health*. Center for Global Development. http://www.cgdev.org/publication/moral-imperative-toward-cost-effectiveness--global-health. Accessed 13 Jan 2015.

Orszag, P.R., and P. Ellis. 2007. Addressing rising health care costs–A view from the congressional budget office. *The New England Journal of Medicine* 357(19): 1885–1887.

Orszag, P.R., and E.J. Emanuel. 2010. Health care reform and cost control. *The New England Journal of Medicine* 363(7): 601–603.

Palmer, S., and J. Raftery. 1999. Economic notes: Opportunity cost. *BMJ* 318(7197): 1551–1552.

Paltiel, A.D., and H.A. Pollack. 2010. Price, performance, and the FDA approval process: The example of home HIV testing. *Medical Decision Making* 30(2): 217–223.

Prentice, E.D., D.A. Crouse, and M.D. Mann. 1992. Scientific merit review: The role of the IACUC. *ILAR Journal* 34(1–2): 15–19.

Resnik, D.B., and A.E. Shamoo. 2005. Bioterrorism and the responsible conduct of biomedical research. *Drug Development Research* 63(3): 121–133.

Rettig, R.A. 1994. Medical innovation duels cost containment. *Health Affairs* 13(3): 7–27.

Rid, A., and D. Wendler. 2010. Risk-benefit assessment in medical research–critical review and open questions. *Law, Probability & Risk* 9(3–4): 151–177.

Roughead, E.E., A.L. Gilbert, and A.I. Vitry. 2008. The Australian funding debate on quadrivalent HPV vaccine: A case study for the national pharmaceutical policy. *Health Policy* 88(2–3): 250–257.

Rudan, I. 2012. Global health research priorities: Mobilizing the developing world. *Public Health* 126(3): 237–240.

Schrecker, T. 2013. Interrogating scarcity: How to think about "resource-scarce settings". *Health Policy and Planning* 28(4): 400–409.

Schüklenk, U., and R.E. Ashcroft. 2002. Affordable access to essential medication in developing countries: Conflicts between ethical and economic imperatives. *Journal of Medicine and Philosophy* 27(2): 179–195.

Shah, S.K. 2013. Outsourcing ethical obligations: Should the revised common rule address the responsibilities of investigators and sponsors? *The Journal of Law, Medicine & Ethics* 41(2): 397–410.

Spinello, R.A. 1992. Ethics, pricing and the pharmaceutical industry. *Journal of Business Ethics* 11(8): 617–626.

Thomson, S., L. Schang, and M.E. Chernew. 2013. Value-based cost sharing in the United States and elsewhere Can increase patients' use of high-value goods and services. *Health Affairs* 32(4): 704–712.

Tobert, J.A. 2003. Case history: Lovastatin and beyond: The history of the HMG-CoA reductase inhibitors. *Nature Reviews Drug Discovery* 2(7): 517–526.

Torgerson, D.J., and S. Byford. 2002. Economic modelling before clinical trials. *BMJ* 325(7355): 98.

Ubel, P.A., and R.M. Arnold. 1995. The unbearable rightness of bedside rationing. Physician duties in a climate of cost containment. *Archives of Internal Medicine* 155(17): 1837–1842.

Weinberger, S.E. 2011. Providing high-value, cost-conscious care: A critical seventh general competency for physicians. *Annals of Internal Medicine* 155(6): 386–388.

Wendler, D., E.J. Emanuel, and R.K. Lie. 2004. The standard of care debate: Can research in developing countries be both ethical and responsive to those countries' health needs? *American Journal of Public Health* 94(6): 923–928.

Wilensky, G.R. 2006. Developing a center for comparative effectiveness information. *Health Affairs* 25(6): w572–w585.

Woods, S., and K. Taylor. 2008. Ethical and governance challenges in human fetal tissue research. *Clinical Ethics* 3(1): 14–19.

Yamey, G. 2002. The world's most neglected diseases: Ignored by the pharmaceutical industry and by public-private partnerships. *BMJ* 325(7357): 176.

Zywicki, T.J. 2007. Institutional review boards as academic bureaucracies: An economic and experiential analysis. *Northwestern University Law Review* 101(2): 861–896.

Chapter 3
From Altruists to Workers: What Claims Should Healthy Participants in Phase I Trials Have Against Trial Employers?

Rebecca A. Johnson

Abstract Phase I trials, which test the safety and toxicity of an investigational agent, are a vital stage of drug development. Many of these trials enroll healthy participants and recent data suggest that some of the healthy participants treat phase I research participation as a form of work. This chapter examines three facets of the shift from research participation as a form of altruism to research participation as a form of work. First, I set out three features of trial participation that support labeling healthy participants' enrollment in phase I research as a form of work. Second, I ask: is phase I research participation similar to risky occupations such as firefighting or coal mining, or is phase I research participation similar to non-risky, low-wage occupations such as janitorial work? To answer this question, I draw upon original data from a systematic review of 475 phase I trials with healthy participants that measures the risk level of the trials. Third, once I have found the appropriate "occupational bucket" for phase I work, I briefly examine the implications for contested questions within research ethics, such as the information persons need prior to consent, rights of withdrawal and compensation for injury, and efforts to increase the transparency of trial results. I argue that conceiving of phase I research as a form of work can bolster the rights of research participants in some of these areas and that bioethicists ought to be less wary of this shift in research participants' roles.

3.1 Introduction

Phase I trials, which test the safety and toxicity of an investigational agent, are a unique form of clinical research. While later phases of clinical trials enroll ill persons and tests substances and interventions for the person's disease area, phase I research often enrolls healthy participants who lack any disease (Wachbroit 2010). The rationales for this enrollment range from *scientific* (drug-related adverse events are better isolated in persons who lack disease symptomatology) to *economic* (for

---

R.A. Johnson (✉)
Princeton University, 117 Wallace Hall, Princeton 08540, NJ, USA
e-mail: raj2@princeton.edu

most diseases, there is a much larger pool of persons who *lack* the disease than who have the disease, creating potential efficiency gains in recruitment) to *ethical* (while randomizing a person with a disease to a placebo arm can be ethically problematic, healthy participants *prefer* placement in the placebo arm). In a previous era, these healthy persons were often recruited from the ranks of prisoners, who were valued both for their abundance and for the guarantee that they would "show up" (Hornblum 1997). When changes to regulations effectively banned recruiting prisoners as healthy participants, trial sponsors moved to recruiting healthy persons from the general population. At present, most commentators accept the practice of enrolling healthy persons from the general population in phase I research as ethically justified.

Despite the general acceptance of phase I research with healthy participants, doubts remain. Most doubts center on the fact that phase I trials expose healthy persons to risk for the benefit of others and on the assumption that phase I trials, some of which are "first-in-human" tests of medicine, are highly risky. Therefore, phase I trials may exploit healthy persons by exposing them to high burdens with no health-related benefits (Elliott and Abadie 2008). The idea that phase I research poses high risks to healthy persons has led some to question the moral permissibility of enrolling healthy persons at all (Wachbroit 2010), others to worry that payment leads participants to ignore or underestimate the significant risks they face (McNeill 1997), and still others to compare the act of participation in phase I trials to hazardous occupations such as coal mining, asbestos removal, firefighting, or soldiering (Abadie 2010; Halpern 2011; Jones and Liddell 2009; Miller and Wertheimer 2007; Resnik 2001; Siminoff 2001).

The present chapter focuses on this last claim: that phase I research is similar to hazardous occupations such as coal mining, firefighting, or soldiering. It decomposes the claim into two parts. First, can we rightfully consider phase I research to be a form of work for the healthy persons who participate? If so, the analogies commentators make to hazardous occupations are coherent. Section 3.2 addresses this question. Second, if we *can* rightfully consider phase I research to be a form of work, what sort of work is it comparable to in terms of ethically relevant characteristics, such as its level of risk? Section 3.3 addresses this question. The reason these questions matter is because many debates about the ethical claims phase I participants have against trial sponsors—for example, should there be a ceiling on the amount of compensation a phase I participant receives?—seem to hinge on the *nature of the relationship* between participants and trial sponsors. If healthy participants in phase I research are altruistic volunteers, comparable to healthy persons who volunteer in hospitals trying to cheer up sick patients, this relationship gives rise to a different set of claims against trial sponsors than if healthy participants in phase I research are workers, comparable to healthy persons who work as coal miners.

Therefore, having determined both *if* phase I research is a form of work for healthy persons and if so, *what* sort of occupation it is, Sect. 3.4 explores two sets of implications of these findings for research ethics. In sum, the aim of this chapter is to explore if there has been a shift from research participation as a form of altruistic volunteering to research participation as a form of work, and the implications of this shift for the obligations trial sponsors have to trial participants.

3.2 Is Phase I Research a Form of Work?

When do a cluster of activities constitute a form of work? In this section, I propose three definitions of work that can help answer this question.[1] In the style of conceptual analysis, these three ways of characterizing activities as work are reconstructed from ideas implicit in our current cultural understanding of work and are meant to provide three *social definitions* of what counts as work.[2] Notably, the *social* definitions of work that I advance are distinct from *legal* definitions of who counts as an employee for the purposes of legally enforced workplace protections.[3] The three social definitions I propose are: an earnings-based conception of work; a time-based conception of work; and a meaning-based conception of work.

3.2.1 Earnings-Based Conception of Work

Take person A and person B. Person A and person B may each cook dinner daily. Person A may *not* consider that cooking to be a form of work (for instance, he may view it as part of his duties in a marriage partnership); person B may consider that cooking to be a form of work. What seems, in part, to distinguish person A and person B's activities is if the person receives (or aims to receive) some form of

---

[1] For the purposes of the present section, I am not making fine-grained distinctions between "work" versus a "job" versus an "occupation" versus a "profession." This is in spite of our colloquial use of the words "occupation" or "profession" to denote higher-status activities than "jobs" or "work." For simplicity's sake, I will consistently use the word "phase I research work."

[2] E.g. Rawls' attempt to find the moral ideas implicit in the background culture of contemporary societies (Rawls 1996, 46).

[3] In a recent book chapter, Holly Fernandez Lynch focuses on whether research participants ought to count as employees (as opposed to volunteers or independent contractors) (Lynch 2014). Because there is a lack of U.S. case law on whether research participants can count as employees, Fernandez Lynch looks to the reasons *why* persons are classified as employees and given certain workplace protections. She outlines two reasons. First, the interests of the employers may diverge from the interests of the employees, meaning that employees need a greater amount of protection for their interests. Second, it is possible to identify an employer party who can be held responsible for the employee's wellbeing. Lynch, analyzing which types of research participation satisfy these two conditions, claims that *most* research participants, rather than solely healthy phase I participants, display these features that the law uses to justify the classification of persons as employees. Her work offers a complementary *legal* definition of work. It helps translate the conceptual arguments I make in the present chapter to concrete legal claims to classify healthy phase I participants as workers under U.S. employment law.

payment in exchange for performing the activities.[4] We can call this the *earnings-based conception of work*: if an activity serves as a substantial source of income for a person, that activity qualifies as work under an *earnings-based conception*.

Phase I research with healthy participants typically counts as work under this definition. Quantitative data from surveys sent to a sample that included both nonprofit academic medical centers and for-profit contract research organizations (CROs) finds that 94 % of the organizations paid persons for study participation, and the organizations often justify this payment as compensation for a person's time (Dickert et al. 2002).[5] The reported wages ranged from $4 to $10 per hour, which, if used to fill a 40 h workweek, ranges from $160 to $400/week (as compared to a minimum-wage worker earning $290/week in the U.S.) (USDOL 2009). Qualitative data from interviews with healthy participants reports a higher estimate: phase I research participants earning up to $1250/week.[6] These data suggest that phase I research has a payment structure similar to wage-based occupations.[7] Since persons are able to financially sustain themselves off of repeated trial participation, phase I research participation qualifies as work under an *earnings-based conception* of work.

3.2.2 Time-Based Conception of Work

An *earnings-based conception of work*, though a useful definition, suffers from some conceptual problems if used as the sole way to analyze whether a certain set of activities counts as work. First is that an earnings-based conception excludes

---

[4] This first conception of work is intertwined with the time-based and meaning-based conceptions of work that I discuss later—that is, it may be morally acceptable to expect someone to perform an activity once for free (for example, helping a friend lift heavy boxes in the moving process) but morally unacceptable to expect a person to perform that activity a large number of times without compensation. Likewise, the act of receiving compensation for an activity may lead that person to view the activity as work (for example, if I am paid to edit the papers of my fellow students I may be more likely to view that editorial work as a job than if I edit those papers for free). Similarly, another characteristic that separates person A and person B in the cooking scenario is if one cooks for the sole benefit of his personal welfare (i.e., cooking daily meals for himself) versus the other cooking to enhance the welfare of others (i.e., he cooks meals for others who visit his restaurant). This distinction is at work in phase I versus phase II/III trials: for phase I trials, healthy persons participate to benefit others with a disease and are often paid for their efforts, versus for later-stage trials, sick persons participate in part to benefit their own disease and are often not paid for their efforts.

[5] Note this includes both trials with healthy persons and with persons with a disease, with organizations reporting little difference in payment policies across the two groups.

[6] For qualitative data on payment, see Abadie (2010).

[7] Unlike the previous criteria, this third criterion is both descriptive—what *is* the payment structure for phase I research participation?—and normative—what *should* be the payment structure for phase I participation? For instance, if someone is participating repeatedly in phase I research activities and thinks of it as a form of work, if they are not currently receiving payment, we can argue that they *should* be receiving payment. For a normative defense of a wage-based payment structure for phase I research participation, see Dickert and Grady (1999) or Shamoo and Resnik (2006).

activities for which people *do not* receive payment but for which persons *should* receive payment. For example, in the U.S., star college football and basketball players generate substantial revenue for their universities but are prohibited from receiving a portion of that revenue as payment. But many argue that these athletes are engaged in work and should receive payment.[8] When making this argument, commentators supplement an *earnings-based conception of work*, which would characterize being a college athlete as not a job because players are not paid, with a *time-based conception of work*. To argue that being a college athlete is a job, commentators point to the fact that the athletes spend an average of 36 h per week devoted to their college athletics, an amount of time in line with the average workweek (Isidore 2014). Therefore, a *time-based conception of work* can help identify activities that *should* be compensated but are not.

The second problem with relying solely on an *earnings-based conception of work* is that there are some activities that provide a substantial source of income to a person but that stretch our commonsense notion of what counts as work. For instance, suppose a person's lottery winnings provide her with a steady source of income for the rest of her life. Yet few persons seem willing to label winning-the-lottery a form of work, and the vast majority of lottery winners (85.5 % in one sample) continue to work after winning (Arvey et al. 2004). This shows that an important part of our commonsense understanding of work is that it occupies the bulk of one's time in addition to provides one with the bulk of one's income.

Turning to this *time-based conception of work*, how does phase I research fare in terms of occupying a substantial portion of a person's time? We can break down time engagement into two features: first, repeat participation in the set of activities and second, one's repeat participation occupying a substantial portion of one's time.

Examining the first criterion—repeat participation—if I starred in one commercial at the age of 13, but as a 25-year old, have done no further acting since then, it would be difficult to claim that I have a job as an actress. Similarly, if persons participate in a single phase I trial—for example, a 2-week inpatient study of an investigational antibody—and do not participate in any phase I studies after that experience, it would be difficult to claim that those persons are phase I research participants by occupation. Examining frequency of phase I trial participation, we see two patterns of participation emerge. The first pattern is *rare* participation—68 % of healthy participants only enroll in 1–2 studies per year (Kass et al. 2007). The second pattern, and the groups relevant for the present paper, is the 22 % who report participating in between three and ten studies and the 3 % who report participating in more than ten studies in 1 year (Kass et al. 2007). These 25 % of people satisfy the repeat participation criterion.

But what about the second criterion—do the three to ten plus studies per year occupy a substantial amount of these participants' time? According to the data described below, the mean duration of study participation is 59 days, with a mode of 7 days that reflects the existence of long phase I trials involving up to a year of engagement. Connecting these two data sources—one on the number of trials a

---

[8] See, e.g., Branch (2011).

participation enrolls into in a given year; another on the duration of those trials—we find that those who participate, for example, in five trials per year may spend between 35 and 295 days engaged in trial participation. Those who participate in eight trials per year may spend between 56 days and a full year engaged in trial participation. So for the 25 % of participants who enroll in three or more trials per year, we can characterize phase I trial participation as a repeat experience rather than continuous activity, and depending on the length of each trial, it is an experience that can occupy a substantial amount of time—approaching or exceeding the 261 days per year that the average full-time U.S. employee spends working (USOPM 2014).[9] Therefore, for this group of participants, phase I trial participation qualifies as work under a *time-based conception of work*. These data show that phase I research participation is more time-consuming than some might expect, and for certain participants, likely qualifies as work under both an *earnings-based* and *time-based* conception.

3.2.3 Meaning-Based Conception of Work

While *time-based* and *earnings-based* conceptions of work are useful definitions, what happens when these definitions offer conflicting guidance on whether an activity is a job? For instance, take the case of college athletes, where an *earnings-based* conception suggests that the activities are not work (since they are uncompensated and considered a form of extracurricular "fun") but a *time-based* conception suggests that the activities are work. A *meaning-based conception of work* can help us resolve this conflict. What meaning does the person engaged in the activity, or a third-party observer speaking about the activity, assign to the activity? Do they label it a form of work?

Examining the meaning assigned to phase I research participation by those who participate and those who conduct the studies, we see several accounts that label phase I research a form of work. A participant interviewed for a large qualitative project on Canadian healthy research participants describes how, "It's a job. ...they were doing a depression study and they were testing some drug... and it was just amazing how people were competing for that job." (Ondrusek 2010). An ethnographer, describing a different group of repeated phase I participants, describes how they thought of the research as follows: "For many participants trials become their full-time job: full-time volunteers might enroll in five to eight trials per year...Some experienced research subjects I met had participated in seventy, eighty...phase I trials over the course of a few years." (Abadie 2010). Yet another study, which interviewed clinical trial coordinators rather than phase I participants themselves, describes the professionalization of phase I research participants into workers who treat the activities as an occupation: "There's kind of a population of professional

---

[9] This number is based on the U.S. national standard for full-time workers of 2087 h per year divided by the standard workday of 8 h.

lab rats. This is kind of what they do, and they don't seem to have any occupations. They do a study every month or two and live off that somehow...It's the repeat business, the 'professionals,' that just do it all the time." (Fisher 2008, 153). These triangulated accounts of phase I research as a set of activities that at least some participants view as a job or a professional identity—one from a set of healthy phase I participants in Canada; another from a set of healthy participants in the eastern United States; still another from a sample of clinical trial coordinators in the southwestern United States—suggests that at least some participants endow their phase I research participation with the meaning of "work."

3.2.4 The Relationship Between Earnings-Based, Time-Based, and Meaning-Based Conceptions of Work

In this section, I outlined three conceptions of work—earnings-based; time-based; meaning-based—and argued that for certain participants, phase I research qualifies as a form of work under each conception. While outlining the precise relationship between the three conceptions of work is beyond the scope of the present chapter, it seems that the more of these characteristics an activity displays, the more defensible it is to characterize the activity as a form of work. Since for certain participants, phase I research displays all three sets of characteristics—it serves as a substantial source of income, occupies a significant proportion of that person's time, and is labeled as work by research participants and trial staff observing those participants—we can firmly characterize it as a form of work for those participants.

3.3 Potentially Problematic Aspects of Phase I Research Work

In the previous section, I established that for at least some subset of healthy participants, phase I research is a form of work under three commonsense understandings of what should count as work—an earnings-based, time-based, and meaning-based conception.[10] In turn, conceiving of phase I research trials as a form of work supports conceiving of the *participants* in phase I research trials in a new way: as workers rather than volunteers. These workers may endorse the goals of biomedical knowledge production and participate in clinical trials as a result of this endorsement. Or, they may view their participation as a job that pays the bills and is easier

---

[10] From this point forward, for shorthand, I will refer to phase I research participation as a form of work as "phase I research work."

to obtain, better compensated, or otherwise preferable to alternative jobs available to them.[11]

Now that I have outlined evidence showing that phase I research participation is a form of work for certain participants, the question that follows is: which aspects of the work should provoke ethical concern? Answering this question matters because different sorts of work display different characteristics that provoke ethical concern. Surrogate pregnancy as a form of work raises concern due to the overlap between the activity in question (pregnancy) and negative stereotypes of females as child bearers first, professionals/persons second. Coal mining as a form of work raises concern due to its physical hazards. Working as a garbage man raises concern due to its status as a "dirty" and therefore status-lowering job (Walzer 1983). Therefore, we need to examine the potentially problematic characteristics of the work in question—is it risky? Is it low status?—to address questions about how to alleviate unfair burdens the work may place upon workers.

We can form this list of potentially problematic characteristics of work by examining our intuitions about what makes phase I research work undesirable and checking those intuitions against data on the actual presence of that undesirable characteristic in phase I research work. To do so, in this section I draw upon original data from a systematic review of phase I research with healthy participants to reject one classification of phase I research work: phase I research as a highly hazardous occupation. I then point to a more appropriate classification: phase I research as an occupation low in flexibility and authority. In Sect. 3.4, I draw out two implications for research ethics of this reclassification of phase I research work.

3.3.1 Risk and Phase I Research Work

Risk initially seems to be the most ethically worrisome aspect of phase I research work. Some compare phase I research participants to "workers performing toxic or dangerous trades such as coal miners, or those exposed to asbestos and other pollutants" (Abadie 2008, 324). Others compare phase I research to "dangerous industries, such as firefighting and mining" (Jones and Liddell 2009). These accounts assume that the most ethically worrisome characteristic of phase I research as a form of work is its level of risk—and equate phase I research work with highly hazardous occupations such as mining, soldiering, and firefighting. Yet these accounts are based on *assumptions* about the level of risk that phase I trials pose to

---

[11] Abadie discusses phase I participants' resistance to the idea that their activities are a form of volunteering instead of a form of work. He also documents the lack of identification with the goals of research among his sample (Abadie 2010).

Others find that among repeat participants, 90 % of repeated healthy participants in phase I research reported that their primary motivation for participating was financial reward. In contrast, medical students who volunteered for phase I research had a more diverse mix of motivations that included a desire to contribute to science in addition to a desire for a financial reward, see Bigorra and Baños (1990).

healthy participants. What is missing from these comparisons of phase I research work to occupations like mining or firefighting is systematic data about the risks that phase I trials pose to healthy participants and how these risks compare to those posed by other occupations.

Drawing upon original data from a systematic review of phase I research trials that enroll healthy participants, I now present this systematic data on the risks of phase I research work. The methods for the systematic review are described in detail elsewhere.[12] In summary, the review analyzed 475 phase I trials that enrolled a total of 27,185 healthy participants. Unlike past reviews of the risks of phase I research, which often focus on a single trial site or a homogeneous trial population, the review of research risks encompasses trials from six continents, testing drugs that target over twenty distinct disease areas, a range of funding sources and trial sites (e.g. contract research organizations hosting pharmaceutical-sponsored trials, academic medical centers hosting foundation-sponsored trials), and a range of different agent types (e.g. biologics, small molecules, and vaccines).

Drawing on risk data from this review and comparing it to data on occupational hazards, I find that coal miners had a 2011 rate of 1.6 work-related deaths per 10,000 fulltime workers, loggers faced 10.4 work-related deaths per 10,000 workers, and trash collectors faced 3.6 work-related deaths per 10,000 workers (USBLS 2013). The trial sample has a rate of study drug-related deaths of 0 per 10,000 "workers." If we look at nonfatal injuries, and define the trials' nonfatal injury rate as the incidence of serious, related adverse events (0.39 per 100 "workers"/year), participants in phase I trials face a lower rate of injuries than muscle strains/sprains for paramedics (2.4 per 100 workers/year) or injury rates for landscapers (2.3 per 100 workers/year). They face similar injury rates to heat burns for food preparers (0.23 per 100 workers/year) or bone fractures for janitors and cleaners (0.21 per 100 workers/year) (USBLS 2010). This suggests that the risks of phase I trials are more similar to the risks of occupations such as food preparers or janitors than the risks of occupations such as coal miners or loggers.

3.3.2 Shifting Phase I Research Work's Occupational Bucket

Does this reclassification of phase I research's occupational bucket—from an occupation similar in risk to coal mining to an occupation similar in risk to janitorial work—mean that there are no remaining ethical concerns regarding *who* seeks work as a phase I trial worker and the negative experiences these persons face on the job? I argue that legitimate ethical concerns remain. And these concerns are related not to the *risks* that phase I research work poses to participants, which my review finds to be rare, but instead to the lack of *occupational flexibility/authority* in phase I research work. While physical hazards that an occupation poses certainly affect

---

[12] Johnson, R.A., A. Rid, E. Emanuel, and D. Wendler. 2016. Risks of phase I research with healthy participants: A systematic review. *Clinical Trials* 13(2): 149–160.

one's wellbeing, additional characteristics of the work seem to also matter for wellbeing. How regimented is the work? Does the work enable persons to build and exercise important physical or intellectual capacities? Does the work leave room for creativity and autonomy? Non-risky work may still fall short on these other wellbeing-affecting dimensions. Therefore, the reclassification of phase I research work from a hazardous occupation to a non-hazardous occupation does not imply that ethical concerns surrounding the work are exhausted.

3.3.3 Occupational Flexibility/Authority and Phase I Research Work

The negative features that occupations such as janitorial work, assembly line work, and phase I research work seem to share are not the hazards that these jobs pose but the regimented nature of the jobs' workflow.[13] The systematic review of phase I research risks presented in this chapter did not specifically examine this regimented nature of phase I research work because sponsors focus on publishing aggregate data about side effects related to the study drug. Sponsors do not explicitly track discomfort that may stem from being confined in an inpatient trial site, having one's freedom restricted by trial requirements such as abstention from caffeine and alcohol, or other characteristics of phase I research work that make it low in *flexibility* (how much discretion does the worker have over which tasks he performs and when he performs them? How much creativity does the job enable?) and the related characteristic of *authority* (how much power is the worker granted to shape the conditions under which he works and to organize the work in such a way that it promotes, rather than sets back, his wellbeing?).[14]

Qualitative research on the subjective experiences of phase I research workers offers vivid examples of the way that phase I research work scores poorly on measures of flexibility and authority.

Phase I research workers report experiencing few physical harms and do not view the potential for physical hazards as the worst aspect of their work. The following quote highlights what they *do* report as the worst aspect of their work: the deprivation of control over their daily activities, which results from study requirements that give participants a strict schedule for what and when to eat, when to sleep, when to lie down, and other closely-regimented aspects of trial participation (Ondrusek 2010, 111–122). As one researcher describes, these participants rate the

---

[13] For a discussion of how work with certain negative aspects can be an appropriate object of liberal distributive justice, see: Arnold (2012) or Hsieh (2008).

[14] Sam Arnold, in his discussion of 'meaningful work' and Rawlsian liberal theory, uses a related package of characteristics to evaluate the quality of an occupation: the authority a worker has, the responsibility a worker is given, and the complexity of the tasks in question. My proposal is inspired by and fully consistent with these criteria for objectively evaluating the quality of an occupation.

deprivation of control as one of the worst aspects of study participation (Ondrusek 2010, 9):

> In some cases subjects are confined to the study facility for periods ranging from several hours to several weeks, where the timing and content of meals are highly regulated and other activities, such as sleep times and levels of physical activity are controlled to meet the requirements of the research protocol. Subjects may, for example, be required to sit up in bed—not lie down and not get up and walk about—for several hours after drug administration. Privacy is limited, as subjects are supervised constantly, even during washroom visits, and must share mealtimes and sleeping quarters.[15]

The structure of phase I research work, which tries to standardize the activities of research participants to a painstaking degree of uniformity, makes the occupation low in flexibility and authority for its workers. In terms of flexibility, the work is terminated if the worker deviates too starkly from the protocol—for example, by missing a dose of the study drug or by refusing a particularly unpleasant research procedure (e.g. a sigmoidoscopy). In terms of authority, phase I research workers argue that they are essentially "paid for passivity"—as one respondent reports, "if you are a guinea pig [phase I research worker] you are enduring something, people are doing things to you and you are just enduring it, you are not actually producing something…I am letting people pay me in exchange for the control they have over me" (Abadie 2006, 94). Some of the relinquishing of control as a phase I research worker is necessary to produce generalizable research knowledge. Yet other losses of control seem arbitrary and rooted in the research staff's desire for authority rather than legitimate scientific reasons.[16] Therefore, although some elements of phase I research work are *inherently* low in flexibility and authority, other low-flexibility elements are rooted in the arbitrary exercise of power by overseeing staff and can be remedied by improving the workplace culture at trial sites.

3.4 Two Implications for Research Ethics

Summing up thus far, in Sect. 3.2, I argued that phase I research, at least for some participants, is a form of work. Section 3.3 asked the question "is phase I research work's level of risk its most problematic feature?"—and drew upon original data from a systematic review of the risks of phase I research to argue that phase I research *cannot* be appropriately characterized as hazardous work. I argued for a re-categorization of phase I research work from "risky work" to "work low in

---

[15] This sentiment is not confined to contemporary descriptions of phase I research work. A brochure that featured the experiences of healthy participants in clinical trials in the 1959 recounts one participant's framing of the negatives of trial participation: "it's the discipline and boredom that get you the most."

[16] For instance, a participant in Ondrusek's study reports how certain research staff would really relish their authority and the ability to act like "drill sergeants" in regulating participant behavior, creating additional frustration at the power imbalance participants experienced between them and the overseeing staff (Ondrusek 2010, 123).

flexibility and authority." In the present section, I briefly draw out the implications of this re-categorization of phase I research work for two debates in research ethics: debates about payment ceilings and debates about the transparency of the data that clinical trials produce. Stakeholders in these two debates often drew on one of two conceptions of healthy phase I research participation: phase I research as a form of altruistic volunteering or phase I research as a hazardous occupation. Since the present chapter rejects both of these characterizations, at least for a subset of phase I research participants, I explore how the new classification of research work lends a new perspective to these two debates.

3.4.1 Payment Ceilings

Two questions arise concerning payment to phase I research workers. First, should these persons be paid at all? And second, if these persons should be paid, should there be a ceiling on the amount of payment they receive?

Addressing the first question, some have argued that offering any payment to participants in phase I research (beyond covering travel expenses) constitutes a form of undue inducement, where persons only consent to participate because of the financial rewards they receive.[17] As others have pointed out, money influences a person's decision to participate in a variety of domains: "most people will not collect garbage, wait tables, change bedpans, pick fruit, work in coal mines, or teach kindergarten, unless they receive some financial or other reward for performing these tasks" (Shamoo and Resnik 2006). Money has an influence, but not necessarily an *undue* influence on a person's decision to perform the work.[18] This is especially the case since data show that the presence of money makes people *more* attuned to risks and other undesirable aspects of the research study in question rather than make people *ignore* or *overlook* these risks (Cryder et al. 2010; Mantzari et al. 2014).

Conceiving of phase I research as a form of work, and indeed, work that may be less hazardous than typical "risky occupations," undermines the claim that phase I research work is so much riskier than other jobs that payment inappropriately lures persons into an exceptionally hazardous situation. One could still argue against payment on other grounds—for example, on the grounds that we ought to have a cultural prohibition against paying persons for work that involves a fairly passive use of one's body, a category that would include both jobs like sex work and jobs like phase I research work; or on the grounds that offering payment may actually lower the quality of phase I research participants by attracting persons who are less

---

[17] Others have also challenged the undue inducement concept, most notably Emanuel (2005).

[18] In the case of work that is clearly immoral—for example, work as a mafia killer—what makes the work immoral are the bad consequences the work has on the wellbeing of others. For these jobs, the immoral aspects are not the level of payment offered but the content of the job itself.

healthy/less ideal test subjects.[19] But arguing against payment on the grounds that phase I research work is much more risky than other jobs is no longer empirically tenable.

Yet among those who argue that *some* payment is warranted, disagreement remains about the *form* that payment should take. Some argue for what I call a *constrained wage* model, where phase I research workers are paid wages comparable to unskilled laborers (but no more) (Dickert and Grady 1999). Others argue that a *constrained wage* model sets unfair ceilings on the payment that phase I research workers receive, and argue for a *free wage* model, where the floor for phase I wages is set by minimum wage regulations but there is no ceiling on compensation (Shamoo and Resnik 2006). Weighing these two models, are there ethical reasons to impose a ceiling on the payment that phase I research workers receive or should the market freely set wages?

Arguments in support of the *constrained wage* model and its payment ceilings fall into two categories: first, some argue that payment ceilings may prevent participants from being inordinately tempted by salaries that increase with the risk level of the trial and second, some people argue that phase I trials should draw people who are motivated by a combination of reasons that includes *both* a desire for payment *and* an endorsement of the socially beneficial goals of research, rather than payment alone (Dickert and Grady 1999).[20] The present chapter shows that risks are rare in phase I research work, undermining the argument for payment ceilings that centers on how high payment tempts people into underestimating risks. It seems less troubling to offer potential participants higher payment in exchange for the unappealingly low levels of flexibility and authority that characterize phase I research work than to offer potential participants higher payments in exchange for high risk. This undermines the first argument supporting a payment ceiling.

With regards to the second argument—payment ceilings will result in participants drawn by money alone rather than participants drawn by a combination of money and altruistic ends—characterizing phase I research participants as workers also undermines this argument. While we may find it preferable if a worker endorses the ends of his employer, we do not generally condone imposing wage ceilings on employment markets to support that goal, especially if the work has unappealing

---

[19] This is a version of the well-known argument that Richard Titmuss makes, where he argues that relying on altruistic volunteers produced a safer blood supply than relying on persons looking for financial incentives (Titmuss 1997). But this argument has been challenged in more recent work by Kieran Healy (2006). With respect to phase I trials, potential participants undergo such extensive testing to ensure that they are "healthy" and not at risk of false adverse events that the problem of financial rewards attracting a less healthy volunteer pool may be minimal. However, more data on the topic are needed.

[20] Another argument that I do not address in the present chapter is that a constrained wage model minimizes the impact of funding disparities between lucrative projects and valuable but non-lucrative research. This seems to be a reality of employment in general—corporate law firms can afford to hire more workers than legal aid bureaus. And it seems best corrected at the governmental level—e.g. trying to incentivize workers to join less lucrative professions through programs like graduate school loan forgiveness for those entering public service—rather than at the level of the individual trial employers making decisions about wages.

aspects. For instance, we would not want to impose a wage ceiling on janitorial work so that the only people drawn to the work are drawn by both money and the social goal of maintaining clean institutions. Likewise, we would not want to impose a wage ceiling on fast food work to draw workers who identify with both the money involved and the goals of corporate fast food production. Drawing on these analogies, the present chapter's argument that phase I research is a form of work with certain unappealing aspects seems to imply that a *free wage* model that establishes a minimum wage but places no cap on payment is normatively preferable than a *constrained wage* model that sets a ceiling on payment.

3.4.2 Trial Data Transparency

The second implication of the present chapter's argument, and an implication less explored than debates about payment for phase I research workers, involves debates about clinical trial data transparency. Phase I research workers help produce data about investigational medical products. Should we impose moral expectations upon trial sponsors to publish the data from all, and not just selected, clinical trials? Should this data go beyond reporting aggregate rates of adverse events to report participant-level harms? Addressing these questions, many, frustrated with the present status of poor data sharing in medical research, have argued for a greater degree of transparency in clinical trial data (Doshi et al. 2013; Lundh et al. 2011; Mello et al. 2013; Zarin 2013). Some base these arguments on the ground that withholding data is unfair to the participants who altruistically volunteer for the trials that produce that data. For instance, Deborah Zarin argues for participant-level data sharing on the grounds that "medical progress is only possible because *altruistic volunteers* put themselves at risk in clinical trials" (emphasis added), and that participant-level reporting serves the "ultimate goal of honoring each trial volunteer's altruism" (Zarin 2013). The argument by commentators like Zarin seems to be as follows:

Premise 1	All persons who participate in phase I clinical trials do so for the altruistic reason of wanting to help disease sufferers.
Premise 2	Increased data transparency helps disease sufferers more than data opacity.
Conclusion	Increased data transparency best honors the reason of persons wanting to participate in clinical trials.

The present chapter challenges premise 1 for *some subset*, though certainly not all, of healthy phase I research participants. These persons may be more concerned with the payment they receive from the trial than how much the results of the trial help disease sufferers. Indeed, some participants are outspoken regarding their suspicion about the value of certain clinical trials for disease sufferers.[21] Data docu-

---

[21] As one participant in Abadie's study reports, "I am pretty cynical and don't think that the trials result in much medical benefit and most of the guinea pigs feel the same way" (Abadie 2010).

menting these views is drawn from qualitative studies, so it is impossible to identify the breakdown of persons who participate in phase I trials for altruistic reasons, and who would likely support trial data transparency, versus persons who participate for other reasons and are either agnostic about or actively opposed to increased between-company sharing of trial data.

Yet the presence of at least some subset of healthy participants who violate premise one of the above argument suggests that it may be better for proponents of increased transparency to argue for data sharing on other grounds, many of which are convincing: basing clinical practice guidelines on biased research evidence poses risks to consumers; the scientific process is more trustworthy if persons impose multiple checks against data fraud or manipulation. These paths to arguing for increased data sharing may be more promising than the argument that healthy phase I participants, because of their uniformly altruistic reasons for participating, would support increased clinical trial data sharing if given a say in the matter.

3.5 Conclusion

This chapter offers a new account of research with healthy participants. I argue that at least for some subset of participants, phase I research is a form of work. Yet in contrast to past accounts, which analogize phase I research work to hazardous occupations such as coal mining or firefighting, I draw upon original data for a new categorization of phase I research work: an occupation low in flexibility and authority. Previous conceptions argued that phase I research work's high risks made it an object of ethical concern. I have argued that phase I research work's highly regimented structure, in which participants have little control and discretion over their daily tasks, is a more appropriate object of ethical concern.

In doing so, I try to counter research exceptionalism by highlighting similarities between the characteristics of phase I research participation and the characteristics of other forms of work (Miller and Wertheimer 2007; Wertheimer 2011). To the extent that commentators propose exceptional measures for phase I research—for instance, ceilings on payment or more rigorous data sharing requirements than other industries mandate—these measures should be justified on grounds other than the misplaced idea that phase I research is *exclusively* composed of altruistic volunteers rather than a mix between altruistic volunteers and more self-interested workers.[22] The chapter also highlights a new focus for debates about phase I research participa-

---

[22] As pointed out in the previous section, because most existing studies of healthy phase I participants have been qualitative rather than quantitative, we do not know the breakdown of altruistic volunteers versus more self-interested workers. As a result, the present chapter tries to rebut arguments that rest on the notion that *all* phase I participants contribute to the trials for altruistic reasons, but still leaves room for others (1) to show empirically that most phase I participants contribute for altruistic reasons, and (2) to argue normatively that the fact that these altruists make up the majority of phase I participants means that ethical analyses should center around these altruistic participants rather than the subset of more "worker-like" participants.

tion. While commentators should continue to scrutinize the risks of phase I trials, they should also investigate the loss of control that participants experience, the arbitrary deprivation of authority they may face, and the extent to which highly-regimented research routines are needed to generate important scientific data. Bioethicists can thus move from condemnations of phase I research work to constructive efforts at improving the on-the-job experience of "professional guinea pigs."

References

Abadie, R. 2006. *Guinea pig's wage. Risk and commoditization in pharmaceutical research in America* [dissertation]. New York: The City University of New York.

Abadie, R. 2008. A guinea pig's wage: Risk and commoditization in American pharmaceutical research. In *Killer commodities: Public health and the corporate production of harm*, ed. M. Baer and H. Baer, 311–334. Manham: Rowman & Littlefield.

Abadie, R. 2010. *The professional guinea pig: Big pharma and the risky world of human subjects*. Durham: Duke University Press.

Arnold, S. 2012. The difference principle at work. *Journal of Political Philosophy* 20(1): 94–118.

Arvey, R.D., I. Harpaz, and L. Hui. 2004. Work centrality and post-award work behavior of lottery winners. *The Journal of Psychology* 138(5): 404–420.

Bigorra, J., and J.E. Baños. 1990. Weight of financial reward in the decision by medical students and experienced healthy volunteers to participate in clinical trials. *European Journal of Clinical Pharmacology* 38(5): 443–446.

Branch, T. 2011. *The shame of college sports*. The Atlantic. 2011 Sept 7. http://www.theatlantic.com/magazine/archive/2011/10/the-shame-of-college-sports/308643. Accessed 25 Dec 2014.

Cryder, C.E., A.J. London, K.G. Volpp, et al. 2010. Informative inducement: Study payment as a signal of risk. *Social Science & Medicine* 70: 455–464.

Dickert, N., and C. Grady. 1999. What's the price of a research subject? Approaches to payment for research participation. *New England Journal of Medicine* 341(3): 198–203.

Dickert, N., E. Emanuel, and C. Grady. 2002. Paying research subjects: An analysis of current policies. *Annals of Internal Medicine* 136(5): 368–373.

Doshi, P., K. Dickersin, D. Healy, et al. 2013. Restoring invisible and abandoned trials: A call for people to publish the findings. *BMJ* 346: f2865.

Elliott, C., and R. Abadie. 2008. Exploiting a research underclass in phase 1 clinical trials. *New England Journal of Medicine* 358(22): 2316–2317.

Emanuel, E.J. 2005. Undue inducement: Nonsense on stilts? *The American Journal of Bioethics* 5(5): 9–13.

Fisher, J. 2008. *Medical research for hire: The political economy of pharmaceutical clinical trials*. New Brunswick: Rutgers University Press.

Halpern, S.D. 2011. Financial incentives for research participation: Empirical questions, available answers and the burden of further proof. *The American Journal of the Medical Sciences* 342(4): 290–293.

Healy, K. 2006. *Last best gifts: Altruism and the market for human blood and organs*. Chicago: University of Chicago Press.

Hornblum, A.M. 1997. They were cheap and available: Prisoners as research subjects in twentieth century America. *BMJ* 315: 1437–1441.

Hsieh, N. 2008. Survey article: Justice in production. *Journal of Political Philosophy* 16(1): 72–100.

Isidore C. 2014. *Playing college sports: A long, tough job.* CNN Money. 2014 Mar 31. http://money.cnn.com/2014/03/31/news/companies/college-athletes-jobs. Accessed 25 Dec 2014.

Johnson, R.A., A. Rid, E. Emanuel, and D. Wendler. 2016. Risks of phase I research with healthy participants: A systematic review. *Clinical Trials* 13(2): 149–160.

Jones, E., and K. Liddell. 2009. Should healthy volunteers in clinical trials be paid according to risk? Yes. *BMJ (Clinical Research Ed.)* 339: b4142.

Kass, N.E., R. Myers, E.J. Fuchs, et al. 2007. Balancing justice and autonomy in clinical research with healthy volunteers. *Clinical Pharmacology and Therapeutics* 82(2): 219–227.

Lundh, A., L.T. Krogsbøll, and P.C. Gøtzsche. 2011. Access to data in industry-sponsored trials. *Lancet* 378(9808): 1995–1996.

Lynch, H.F. 2014. Protecting human research subjects as human research workers. In *Human subjects research regulation*, ed. I.G. Cohen and H.F. Lynch, 327–340. Cambridge: MIT Press.

Mantzari, E., F. Vogt, and T. Marteau. 2014. Does incentivizing pill-taking 'crowd out' risk-information processing? Evidence from a web-based experiment. *Social Science and Medicine* 106: 75–82.

McNeill, P. 1997. Paying people to participate in research: Why not? *Bioethics* 11(5): 390–396.

Mello, M.M., J.K. Francer, M. Wilenzick, et al. 2013. Preparing for responsible sharing of clinical trial data. *New England Journal of Medicine* 369(17): 1651–1658.

Miller, F.G., and A. Wertheimer. 2007. Facing up to paternalism in research ethics. *Hastings Center Report* 37(3): 24–34.

Ondrusek, N. 2010. *Making participation work: A grounded theory describing participation in phase I drug trials from the perspective of the healthy subject* [dissertation]. Toronto: University of Toronto.

Rawls, J. 1996. *Political liberalism*. New York: Columbia University Press.

Resnik, D.B. 2001. Research participation and financial inducements. *American Journal of Bioethics* 1(2): 54–56.

Shamoo, A.E., and D.B. Resnik. 2006. Strategies to minimize risks and exploitation in phase one trials on healthy subjects. *American Journal of Bioethics* 6(3): W1–W13.

Siminoff, L.A. 2001. Money and the research subject: A comment on Grady. *American Journal of Bioethics* 1(2): 65–66.

Titmuss, R. 1997. *The gift relationship: From human blood to social policy*. London: LSE Books.

United States Bureau of Labor Statistics (USBLS). 2010. *Incidence rates of nonfatal occupational injuries and illnesses requiring days way from work.* http://www.bls.gov/iif/oshwc/osh/case/ostb2829.pdf. Accessed 25 Dec 2014.

United States Bureau of Labor Statistics (USBLS). 2013. *Census of fatal occupational injuries.* http://www.bls.gov/iif/oshwc/cfoi/cfch0010.pdf. Accessed 25 Dec 2014.

United States Department of Labor (USDOL). 2009. *Minimum wage.* http://www.dol.gov/dol/topic/wages/minimumwage.htm. Accessed 25 Dec 2014.

United States Office of Personnel Management (USOPM). 2014. *Computing hourly rates of pay using the 2,087-hour divisor.* http://www.opm.gov/policy-data-oversight/pay-leave/pay-administration/fact-sheets/computing-hourly-rates-of-pay-using-the-2087-hour-divisor. Accessed 25 Dec 2014.

Wachbroit, R. 2010. Assessing phase I clinical trials. *Law, Probability and Risk* 9(3–4): 179–186.

Walzer, M. 1983. *Spheres of justice: A defense of pluralism and equality*. New York: Basic Books.

Wertheimer, A. 2011. *Rethinking the ethics of clinical research: Widening the lens*. Oxford: Oxford University Press.

Zarin, D.A. 2013. Participant-level data and the new frontier in trial transparency. *New England Journal of Medicine* 369(5): 468–469.

Chapter 4
Nocebo Effects: The Dilemma of Disclosing Adverse Events

Luana Colloca

Abstract Any randomised clinical trial (RCT) is characterised by the emergence of adverse events. Some adverse events are related to the action of the active drug, but a substantial proportion is due to patients being alerted to potential adverse events as part of the informed consent process. Presenting patients with side effects of treatments and interventions can induce so-called "nocebo effects", which refers to adverse events related to negative expectations and anticipations. Neurobiological and pre-clinical studies have shown that nocebo effects result from negative expectations, previous experiences, and clinical encounters. Disclosures and the manner in which information is delivered can contribute to producing adverse effects. This phenomenon poses an ethical conundrum as the patient has the right to be informed about potential risks and side effects of a treatment, yet a detailed disclosure may produce undesirable harm. We discuss state-of-the art nocebo research and associated clinical and ethical implications.

4.1 Introduction

The term "nocebo" refers to the deterioration of outcomes due to negative expectations and represents negative "placebo" effects (Kennedy 1961; Kissel 1964). For decades, these observations were dismissed as purely psychological effects. Current research indicates that nocebos can cause real biological changes, a finding that may transform how patient-doctor communication is framed and practiced. Verbal communication, providers' behaviour, environmental cues, and the appearance of medical devices may induce negative expectations that lead to adverse effects in

---

L. Colloca (✉)
Department of Pain and Translational Symptom Science, School of Nursing,
University of Maryland Baltimore, Baltimore, MD, USA

Department of Anesthesiology, School of Medicine, University of Maryland Baltimore,
Baltimore, MD, USA
e-mail: colloca@son.umaryland.edu

both research subjects and patients. Like placebo effects, nocebo effects can strongly increase outcomes across different clinical conditions (Benedetti et al. 2007; Colloca and Miller 2011).

Negative expectations can be created through anticipation of worsening via verbal suggestions or prior exposure to negative symptoms. Importantly, negative expectations produce nocebo effects that are comparable in magnitude to those induced by actual experience of increases in somatosensory perception (Colloca et al. 2008), pain (Colloca et al. 2008; van Laarhoven et al. 2011), itching (van Laarhoven et al. 2011) and worsening of motor performance (Pollo et al. 2012).

Vicarious learning represents another mechanism involved in the formation of nocebo effects, since nocebo effects can also be induced by observing other people in pain (Colloca and Benedetti 2009; Vogtle et al. 2013). Observationally-induced nocebo effects can also account for mass psychogenic illness (Mazzoni et al. 2010). In the study by Mazzoni et al., research subjects were asked to inhale a sample of normal air but were told that it contained a suspected environmental toxin known to cause headache, nausea, itchy skin and drowsiness. Half of the participants observed an actor inhale and display the four expected symptoms. Participants who observed another person become ill displayed significant increased reports of expected symptoms (Mazzoni et al. 2010). These findings suggest the importance of social learning in shaping nocebo effects in mass psychogenic illness models with potential implications for public health (Hahn 1997).

Negative expectations can also influence drug outcomes often in a paradoxical manner. For example, asthmatic patients reported bronchodilatation as a response to bronchoconstrictors described as bronchodilators, and vice versa bronchoconstriction when bronchodilatators were presented as bronchoconstrictors (Luparello et al. 1970). Similarly, healthy subjects who believed that they were given a stimulant, perceived an increase of their muscle tension when they were actually receiving a muscle relaxant medication (Flaten et al. 1999).

Nocebo effects can significantly increase nonspecific symptoms and complaints in patient populations, resulting in psychological distress, medication nonadherence, and need for additional medicines to treat the nocebo adverse effects (Barsky et al. 2002). For example, headaches, which are a common side effect of antidepressants, can result simply from the mention of headaches in the informed consent process as a potential side effect. Indeed, in randomised controlled trials (RCTs), a significant proportion of depressed patients who received placebos reported headache (Mora et al. 2011; Rief et al. 2011). Amanzio and co-workers performed a systematic review of adverse effects of anti-migraine randomised placebo-controlled clinical trials (Amanzio et al. 2009). The final sample consisted of 69 studies including 56 trials for triptans, 9 trials for anticonvulsants and 8 trials for non-steroidal anti-inflammatory drugs (NSAIDs). The authors found a high rate of adverse events in the placebo arms of trials with anti-migraine drugs matching those described for real drugs. For example, anticonvulsant placebos produced anorexia, memory difficulties, paraesthesia and upper respiratory tract infection—all adverse events reported in the side effect profile of the three classes of anti-migraine drugs (Amanzio et al. 2009).

The link between reported side effects in the placebo groups and the known side effects of particular drugs suggests genuine nocebo effects from the informed consent process. It is important to clarify that these effects can represent either an *apparent* or a *true* nocebo effect. If patients report the same prevalence of headaches in a no-treatment control group that did not receive placebos, it is likely that the adverse event represents merely an *apparent* nocebo effect. Hence, the side effects observed in the placebo group may reflect the natural history of the condition or common symptoms that everyone experiences, rather than true nocebo effects. From a methodological viewpoint, nocebo effects should be factored in RCTs by either including a no-treatment group that does not receive placebos or by including a group that is not informed about the side effects related to a treatment under investigation. However, both of these alternatives present ethical constraints because giving a placebo may not be feasible since intentional concealment of the information violates the patient's rights (Colloca and Miller 2011).

4.2 Nocebo Effects and Lack of Adherence

Discontinuation and lack of adherence are also common problems in RCTs and practice, mostly related to the occurrence of both adverse events and nocebo effects. Symptoms such as restlessness, nausea, anorexia and insomnia have been reported in the placebo arms of RCTs investigating fatigue in patients with advanced cancer (e.g. breast cancer) (de la Cruz et al. 2010). In particular, nausea, one of the most debilitating and severe side effects in cancer patients, is strongly modulated by nocebo effects negatively affecting nutrition, adherence to therapy and quality of life (Colagiuri and Zachariae 2010; Stockhorst et al. 1998).

In RCTs, communication of adverse effects often leads to withdrawal from the study. For example, Myers et al. studied the effect of mentioning gastrointestinal side effects during the consent process in a randomised, double-blind, placebo-controlled trial examining the benefit of either aspirin, sulphinpyrazone, or both drugs, for unstable angina pectoris. They found that the inclusion of potential gastrointestinal side effects in the informed consent forms resulted in a sixfold increase in gastrointestinal symptoms with consequent patient-initiated cessation of therapy (Cairns et al. 1985; Myers et al. 1987).

Nocebo effects produce discontinuation and lack of adherence also in RCTs for statin drugs in population-based studies. In statin trials performed between 1994 and 2003 placebo groups presented a variety of symptoms such as headache ranging from 0.2 to 2.7 %, and abdominal pain from 0.9 to 3.9 %. Interestingly, the adverse effects observed in the general population were higher than those found in clinical trials of statin drugs (Rief et al. 2006).

In the field of pain medicine, nocebo responses are relevant and produce dropouts and harms (Mitsikostas et al. 2011, 2012; Papadopoulos and Mitsikostas 2012) (Table 4.1).

Table 4.1 Nocebo responses in the arena of pain diseases

Disease	Treatment	Nocebo responses (%)	Drop-out (%)	Ref.
Migraine	Symptomatic treatments	18.45	0.33	Mitsikostas DD et al. Cephalalgia. (2011)
	Preventive treatments	42.78	4.75	
Tension-type headache	Preventive treatments	23.99	5.44	Mitsikostas DD et al. Cephalalgia. (2011)
Neuropathic pain	Symptomatic treatments	52	6	Papadopoulos J Neurol. (2012)
Fibromyalgia	Symptomatic treatments	67.20	9.50	Mitsikostas DD et al. Eur J Neurol. (2012)

For example, the proportion of nocebo responses in RCTs of symptomatic treatments for migraine is about 18 % with dropouts of 0.33 % (Mitsikostas et al. 2011). This percentage becomes even higher for neuropathic pain RCTs with 52 % of nocebo responses and a dropout of 6 % (Papadopoulos and Mitsikostas 2012) and still more for fibromyalgia RCTs with 67.2 % of nocebo responses and 9.50 % dropouts (Mitsikostas et al. 2012).

4.3 Nocebo Effects and Framing Effects

In order to better assess the relation between framing effects during informed consent processes or interventional procedures and the occurrence of nocebo effects, preclinical studies have been performed with different versions of the contents of disclosures. These studies show that informing patients about potential adverse effects of a specific treatment, elicited drug-like adverse events in patients with allergic disorders, Parkinson's disease, anxiety, pain, and sexual disorders. For example, in a double-blind study of symptom provocation, a quarter of patients with food allergies developed allergic symptoms when injected with saline that was described to them as an allergen (Jewett et al. 1990). Eighteen patients were tested in 20 sessions by the same technician, using the same extracts at the same dilutions with the same saline diluent. In each session three injections of extract and nine of diluent were given in random sequence. The symptoms evaluated included nasal stuffiness, dry mouth, nausea, fatigue, headache, and feelings of disorientation or depression. The responses of the patients to the active and control injections were indistinguishable, as was the incidence of positive responses. When the provocation of symptoms to identify food sensitivities was evaluated under double-blind conditions, the frequency of positive responses to the injected extracts appeared to be the result of suggestion and chance (Jewett et al. 1990). Moreover, outpatients with adverse drug reactions (ADR) undergoing oral drug challenges presented both subjective symptoms such as itching, nausea, headache, and abdominal pain when they actually received a placebo. The nocebo effects were not limited to subjective symptoms but nocebo effects influenced objective symptoms as well as dyspnoea, cough, hypotension, tachicadia, erythaema, and urticaria (Liccardi et al. 2004; Lombardi et al. 2008).

Patients with Parkinson's disease presented a worsening of bradykinaesia referring to extreme slowness of movements and reflexes, if they were told that the device implanted in their brains to deliver high frequency stimulation to the subthalamic nuclei was turned off when in actuality it was active (Benedetti et al. 2003; Colloca et al. 2004). Information was intentionally manipulated to explore the effects of information also in patients treated for post-operative pain and anxiety. Patients openly informed about the interruption of treatment experienced a sudden increase of anxiety and pain, whilst a hidden interruption (controlled by computer) did not induce a deterioration, thus, suggesting that the mere communication of treatment interruption aggravated patients' symptoms (Colloca et al. 2004).

Communicating about adverse side effects also induced nocebo effects in sexual disorders. Patients with benign prostatic hyperplasia (BPH) received finasteride (5 mg) described as a "compound of proven efficacy for the treatment of BPH" and were randomised to two different disclosure groups. One group was informed that the medication "…may cause erectile dysfunction, decreased libido, problems of ejaculation but these are uncommon"; the other group was not told about these side effects. A 6- and 12-month follow-up showed that finasteride administration produced a significantly higher rate of reported sexual side effects in those patients who were informed about the possibility of sexual dysfunction (43.6 %) as compared to those in whom the same information was omitted (15.3 %) (Mondaini et al. 2007). Although concealment of adverse events is problematic in daily clinical practice, this study suggests that the therapeutic effects of patient-clinician communication are not limited in order to motivate patients to adhere to a recommended treatment regimen, to choose a healthier lifestyle, to adopt better psychological attitudes, but also to avoid occurrence of nocebo effects (Miller and Colloca 2011).

Patient-clinician communication and framing effects can promote beneficial placebo effects and minimise nocebo reactions to pain. Women at term gestation requesting epidural analgesia were randomised to one of two descriptions of the pain experience during the epidural procedure (Varelmann et al. 2010). Participants were randomised to two disclosures: (1). "You are going to feel a big bee sting; this is the worst part of the procedure"; and (2). "We are going to give you a local anesthetic that will numb the area and you will be comfortable during the procedure". The first description reflected the standard way to communicate the effect of the procedural intervention while the second description described the procedure anticipating the benefit of the anaesthetic medication. Those women in labour who were told to expect pain like a bee sting during the local anaesthetic injection (nocebo group) rated pain significantly higher than those receiving the procedure along with gentle and positive words (Varelmann et al. 2010). This study emphasised how small changes in the way in which information is framed, impact clinical outcomes, showing that it is possible and ethically acceptable to frame the information in a helpful way in order to prevent nocebo effects and preserve patients' rights to be informed.

4.4 Ethical Considerations

Recent advances in nocebo research outline the need to reconsider the importance of the patient-clinician communication, adverse events induced by negative experiences and expectations in clinical contexts and RCTs, as well as the need to consider the potential of framing effects. It is becoming evident that verbal instructions are powerful in triggering negative expectations with an impact on clinical outcomes. Therefore, a first step is the realisation that a clinician's words and attitudes can potentially facilitate or worsen symptoms' improvement and healing processes.

The goal standard should be to avoid untenable nocebo effects. Information provided along with the administration of active treatments (and placebo used in RCTs) is akin to walking a tightrope of communication. Examples of concealing information (e.g. sexual dysfunction related to taking finasteride for BPH) are debatable, as some patients may not agree to undergo the treatment because of the sexual adverse events. Nevertheless, clinicians as well as researchers performing RCTs, have an obligation to convey truthful information to patients so that they can make informed decisions in light of their personal preferences and values.

According to a so-called "authorised concealment" approach, patients might consent to receive information only about potential serious or irreversible harm. This approach may deserve consideration in circumstances in which the patient is not exposed to serious risks (Colloca and Miller 2011; Miller and Colloca 2011). Nevertheless, the consent process should inform the patient about the concealment of part of the information warning her to report any experienced adverse event promptly. There are at least two potential alternatives to the "authorised concealment" approach: (1). Conveying information by taking advantage of framing strategies and, (2). Educating physicians (and scientists running clinical trials) about the reality of the nocebo phenomenon.

A variety of studies have investigated the effects of framing information regarding risks and benefits of interventions on patient decision-making (Edwards et al. 2001), but limited clinically-oriented research has considered the impact of informing patients about the nocebo effects. Examples of choices of framing strategies are apparent in many everyday circumstances. For instance, a physician who is recommending a drug to a patient, may communicate the proportion of patients who experience the side effects. Side effects, such as headaches or nausea, may be mentioned merely as a slight possibility. There is also a choice in communicating the probability of experiencing adverse effects based on extant research, either qualitatively or quantitatively (Peters et al. 2011). Furthermore, this information can be conveyed by focusing on the minority of patients who experience a particular side effect or by focusing on the majority of patients who do not experience the side effect. These different ways of framing side effect information can have a differential impact on patients' perception of adverse events (Woloshin and Schwartz 2011) and potentially, occurrence of nocebo effects. Further research is needed to explore the link between perceived risks and benefits of interventions (Edwards et al. 2001) and nocebo effects.

The second option is to encourage clinician and patient education about nocebo research and its clinical implications (Colloca and Finniss 2012). This perspective is still poorly explored, but clinicians and scientists should be encouraged to systematically tell the patient that some adverse effects occur as a result of informing her about certain side effects. This perspective would require an effort to educate both clinicians and patients about the realm of nocebo research and translate what we have learned in the laboratory settings into daily practices. Importantly, recent surveys of patients showed that patients are open to learning about these phenomena (Hull et al. 2013) thus boosting instead of threatening patient-clinician interactions. Informing patients about the possibility of experiencing nocebo effects is consistent with the benefit of incorporating framing strategies to minimize nocebo effects while informed consent and respect for the patient's autonomy are guaranteed. Trialists should also consider nocebo reactions and the link between conveying information and occurrence of certain negative outcomes. Surprisingly, although Walter Kennedy in his 1961 article, "The nocebo reaction" indicated the possibility that useful drugs have been often discarded because of an "appreciable number of nocebo reactors in the test subjects" (Kennedy 1961), trialists have yet to heed Walter Kennedy's thoughts.

In conclusion, nocebo research provides evidence supporting the claim that patient-clinician communication has effects on clinical outcomes. Therefore, concerns about trustfulness should not impede helpfulness and pragmatism, which are two key morally relevant aspects guiding clinicians and patients routinely in therapeutic decision making processes.

Acknowledgment This research is supported by the National Institute of Dental and Craniofacial Research (1R01DE025946-01, LC) and the University of Maryland, Baltimore, Department of Pain and Translational Symptom Science, School of Nursing, Baltimore, USA.

References

Amanzio, M., L.L. Corazzini, L. Vase, and F. Benedetti. 2009. A systematic review of adverse events in placebo groups of anti-migraine clinical trials. *Pain* 146(3): 261–269.

Barsky, A.J., R. Saintfort, M.P. Rogers, and J.F. Borus. 2002. Nonspecific medication side effects and the nocebo phenomenon. *JAMA* 287(5): 622–627.

Benedetti, F., A. Pollo, L. Lopiano, M. Lanotte, S. Vighetti, and I. Rainero. 2003. Conscious expectation and unconscious conditioning in analgesic, motor, and hormonal placebo/nocebo responses. *The Journal of Neuroscience* 23(10): 4315–4323.

Benedetti, F., M. Lanotte, L. Lopiano, and L. Colloca. 2007. When words are painful: Unraveling the mechanisms of the nocebo effect. *Neuroscience* 147(2): 260–271.

Cairns, J.A., M. Gent, J. Singer, K.J. Finnie, G.M. Froggatt, D.A. Holder, et al. 1985. Aspirin, sulfinpyrazone, or both in unstable angina. Results of a Canadian multicenter trial. *The New England Journal of Medicine* 313(22): 1369–1375.

Colagiuri, B., and R. Zachariae. 2010. Patient expectancy and post-chemotherapy nausea: A meta-analysis. *Annals of Behavioral Medicine* 40: 3–14.

Colloca, L., and F. Benedetti. 2009. Placebo analgesia induced by social observational learning. *Pain* 144(1–2): 28–34.

Colloca, L., and D. Finniss. 2012. Nocebo effects, patient-clinician communication, and therapeutic outcomes. *JAMA* 307(6): 567–568.

Colloca, L., and F.G. Miller. 2011. The nocebo effect and its relevance for clinical practice. *Psychosomatic Medicine* 73(7): 598–603.

Colloca, L., L. Lopiano, M. Lanotte, and F. Benedetti. 2004. Overt versus covert treatment for pain, anxiety, and Parkinson's disease. *The Lancet Neurology* 3(11): 679–684.

Colloca, L., M. Sigaudo, and F. Benedetti. 2008. The role of learning in nocebo and placebo effects. *Pain* 136(1–2): 211–218.

de la Cruz, M., D. Hui, H.A. Parsons, and E. Bruera. 2010. Placebo and nocebo effects in randomized double-blind clinical trials of agents for the therapy for fatigue in patients with advanced cancer. *Cancer* 116(3): 766–774.

Edwards, A., G. Elwyn, J. Covey, E. Matthews, and R. Pill. 2001. Presenting risk information–a review of the effects of "framing" and other manipulations on patient outcomes. *Journal of Health Communication* 6(1): 61–82.

Flaten, M.A., T. Simonsen, and H. Olsen. 1999. Drug-related information generates placebo and nocebo responses that modify the drug response. *Psychosomatic Medicine* 61(2): 250–255.

Hahn, R.A. 1997. The nocebo phenomenon: Concept, evidence, and implications for public health. *Preventive Medicine* 26(5 Pt 1): 607–611.

Hull, S.C., L. Colloca, A. Avins, N.P. Gordon, C.P. Somkin, T.J. Kaptchuk, et al. 2013. Patients' attitudes about the use of placebo treatments: Telephone survey. *BMJ* 347: f3757.

Jewett, D.L., G. Fein, and M.H. Greenberg. 1990. A double-blind study of symptom provocation to determine food sensitivity. *The New England Journal of Medicine* 323(7): 429–433.

Kennedy, W.P. 1961. The nocebo reaction. *The Medical World* 95: 203–205.

Kissel, P., and D. Barrucand. 1964. *Placebos et effet placebo en medecine*. Paris: Masson.

Liccardi, G., G. Senna, M. Russo, P. Bonadonna, M. Crivellaro, A. Dama, et al. 2004. Evaluation of the nocebo effect during oral challenge in patients with adverse drug reactions. *Journal of Investigational Allergology & Clinical Immunology* 14(2): 104–107.

Lombardi, C., S. Gargioni, G.W. Canonica, and G. Passalacqua. 2008. The nocebo effect during oral challenge in subjects with adverse drug reactions. *European Annals of Allergy and Clinical Immunology* 40(4): 138–141.

Luparello, T.J., N. Leist, C.H. Lourie, and P. Sweet. 1970. The interaction of psychologic stimuli and pharmacologic agents on airway reactivity in asthmatic subjects. *Psychosomatic Medicine* 32(5): 509–513.

Mazzoni, G., L. Foan, M.E. Hyland, and I. Kirsch. 2010. The effects of observation and gender on psychogenic symptoms. *Health Psychology* 29(2): 181–185.

Miller, F.G., and L. Colloca. 2011. The placebo phenomenon and medical ethics: Rethinking the relationship between informed consent and risk-benefit assessment. *Theoretical Medicine and Bioethics* 32(4): 229–243.

Mitsikostas, D.D., L.I. Mantonakis, and N.G. Chalarakis. 2011. Nocebo is the enemy, not placebo. A meta-analysis of reported side effects after placebo treatment in headaches. *Cephalalgia* 31(5): 550–561.

Mitsikostas, D.D., N.G. Chalarakis, L.I. Mantonakis, E.M. Delicha, and P.P. Sfikakis. 2012. Nocebo in fibromyalgia: Meta-analysis of placebo-controlled clinical trials and implications for practice. *European Journal of Neurology* 19(5): 672–680.

Mondaini, N., P. Gontero, G. Giubilei, G. Lombardi, T. Cai, A. Gavazzi, et al. 2007. Finasteride 5 mg and sexual side effects: How many of these are related to a nocebo phenomenon? *The Journal of Sexual Medicine* 4(6): 1708–1712.

Mora, M.S., Y. Nestoriuc, and W. Rief. 2011. Lessons learned from placebo groups in antidepressant trials. *Philosophical Transactions of the Royal Society of London. Series B, Biological Sciences* 366(1572): 1879–1888.

Myers, M.G., J.A. Cairns, and J. Singer. 1987. The consent form as a possible cause of side effects. *Clinical Pharmacology and Therapeutics* 42(3): 250–253.

Papadopoulos, D., and D.D. Mitsikostas. 2012. A meta-analytic approach to estimating nocebo effects in neuropathic pain trials. *Journal of Neurology* 259(3): 436–447.

Peters, E., P.S. Hart, and L. Fraenkel. 2011. Informing patients: The influence of numeracy, framing, and format of side effect information on risk perceptions. *Medical Decision Making* 31(3): 432–436.

Pollo, A., E. Carlino, L. Vase, and F. Benedetti. 2012. Preventing motor training through nocebo suggestions. *European Journal of Applied Physiology* 112(11): 3893–3903.

Rief, W., J. Avorn, and A.J. Barsky. 2006. Medication-attributed adverse effects in placebo groups: Implications for assessment of adverse effects. *Archives of Internal Medicine* 166(2): 155–160.

Rief, W., A.J. Barsky, J.A. Glombiewski, Y. Nestoriuc, H. Glaesmer, and E. Braehler. 2011. Assessing general side effects in clinical trials: Reference data from the general population. *Pharmacoepidemiology and Drug Safety* 20(4): 405–415.

Stockhorst, U., J.A. Wiener, S. Klosterhalfen, W. Klosterhalfen, C. Aul, and H.J. Steingruber. 1998. Effects of overshadowing on conditioned nausea in cancer patients: An experimental study. *Physiology & Behavior* 64(5): 743–753.

van Laarhoven, A.I., M.L. Vogelaar, O.H. Wilder-Smith, P.L. van Riel, P.C. van de Kerkhof, F.W. Kraaimaat, et al. 2011. Induction of nocebo and placebo effects on itch and pain by verbal suggestions. *Pain* 152(7): 1486–1494.

Varelmann, D., C. Pancaro, E.C. Cappiello, and W.R. Camann. 2010. Nocebo-induced hyperalgesia during local anesthetic injection. *Anesthesia and Analgesia* 110(3): 868–870.

Vogtle, E., A. Barke, and B. Kroner-Herwig. 2013. Nocebo hyperalgesia induced by social observational learning. *Pain* 154(8): 1427–1433.

Woloshin, S., and L.M. Schwartz. 2011. Communicating data about the benefits and harms of treatment: A randomized trial. *Annals of Internal Medicine* 155(2): 87–96.

Part II
Challenges in Common Domains of Research Governance

Chapter 5
Discriminating Between Research and Care in Paediatric Oncology—Ethical Appraisal of the ALL-10 and 11 Protocols of the Dutch Childhood Oncology Group (DCOG)

Sara A.S. Dekking, Rieke van der Graaf, Martine C. de Vries, Marc B. Bierings, and Johannes J.M. van Delden

Abstract Paediatric oncology is a classic example of a field in which research and care are closely intertwined. Moreover, bioethicists have argued that in environments such as paediatric oncology we should no longer draw sharp distinctions between research and care. Recently, two Dutch protocols for the treatment of children with Acute Lymphoblastic Leukaemia (ALL) have been categorised in two different ways, one as research (ALL-11) and the other as treatment (ALL-10). We analysed these protocols in order to explore whether the distinction between research and care in paediatric oncology is morally relevant. We applied several characteristics of research to the ALL-10 and 11 protocols: the goal of producing generalisable knowledge; systematic collection of data; potentially high and uncertain risks; burdens and risks unrelated to treatment; and provision of treatment according to detailed protocols. Both ALL-protocols exhibit general characteristics of research. At the same time, both protocols also clearly satisfy the objective of delivering the best available treatment. Therefore, it remains to be discussed how to review these kinds of protocols that integrate a research goal with the objective of providing individual patients with best current treatment. A change in both research ethics regulation and oversight of conventional care is needed. More case studies are essential to expand the moral evaluation of the intertwinement between research and care in paediatric oncology.

S.A.S. Dekking (✉) • R. van der Graaf • M.C. de Vries
M.B. Bierings • J.J.M. van Delden
Department of Medical Humanities, University Medical Center Utrecht,
Julius Center for Health Sciences and Primary Care, 85500,
3508 GA Utrecht, The Netherlands
e-mail: S.A.S.Dekking@umcutrecht.nl

5.1 Introduction

Currently, there is an intensive debate in bioethics about whether the practice of medical research should or should not be strictly distinguished from clinical care (Beauchamp 2011; Brody and Miller 2003; Largent et al. 2011; Weijer and Miller 2003). A classic example of a field in which research and care are highly integrated is paediatric oncology (de Vries et al. 2011; Largent et al. 2011). From the outset paediatric oncology has been constructed as a practice that closely combines research and care (Krueger 2008), in order to overcome the lack of decisive knowledge currently present in paediatric medicine in general (Kimland et al. 2012; Lindell-Osuagwu et al. 2009; Pandolfini and Bonati 2005; 't Jong et al. 2000) and in paediatric oncology in particular (Conroy et al. 2003; van der Berg and Tak 2011). The lack of evidence is due to the fact that research data from adult oncology are not generalisable to children for the most part, and that childhood cancer is a rarity (de Vries et al. 2011). It is estimated that over 70 % of patients take part in clinical trials (Ablett and Pinkerton 2003). Moreover, paediatric oncologists often regard clinical trials as providing state-of-the-art treatment (Joffe and Weeks 2002).

Not just oncologists, but bioethicists have also highlighted the integration of research and care in paediatric oncology. Recently, Kass and colleagues have presented paediatric oncology as an illustration of a practice where research and care are optimally combined for the benefit of both individual patients and groups of patients (Kass et al. 2013). According to Kass et al., the context of paediatric oncology "is constructed to bring the most pertinent forms of scientific understanding to bear on clinical care, and clinical care generates new scientific learning" (Kass et al. 2013, 7). They claim that in environments such as paediatric oncology the distinction between research and care is becoming increasingly blurred and ceases to be of moral importance for determining which activities need ethical oversight This claim is in sharp contrast to the traditional bioethical paradigm that clearly distinguishes medical research from medical care (CIOMS 2002; NCPHSBBR 1979). The distinction is usually based on the premise that research is designed to develop generalisable knowledge for *groups* of patients, whereas the benefits of this knowledge for individual patients participating in the research are uncertain, and care is directly aimed at the promotion of health and wellbeing in *individual* patients (NCPHSBBR 1979; CoE 2005).

Unfortunately, Kass et al. do not provide concrete examples to substantiate their claim about paediatric oncology practice, nor do they pursue its moral implications. We believe that examples from practice can help in gaining insight into the validity and implications of their claims, because such examples provide an empirical assessment that could potentially result in a modification and reformulation of the normative outcome. As such, "theory and practice ... mutually influence each other in the process of searching for reliable moral judgments and theories" (van Thiel and van Delden 2010, 184).

In this article we will explore paediatric oncology treatment protocols that are considered current best treatment, while they are simultaneously designed to answer study questions by means of collecting and evaluating treatment results. Examples of such studies are treatment protocols for children with Acute Lymphoblastic Leukaemia (ALL). Evaluating best available treatments has greatly improved survival rates for children with ALL (Pieters 2010), but simultaneously raises uncertainties as to how these kinds of protocols should be categorised. This ambiguity is illustrated by the ALL-10 and ALL-11 protocols of the Dutch Childhood Oncology Group (DCOG). These two ALL-protocols are largely similar, but have been categorised differently. DCOG ALL-10 has been considered by a Research Ethics Committee (REC), which decided that it was exempt from ethical review. The DCOG ALL-11 protocol has been deemed a research protocol and was reviewed accordingly. We will compare both protocols in order to discover whether the distinction between research and treatment is morally relevant in paediatric oncology.

For our comparison, we will use the five "characteristics" of research that Kass et al. have recently listed as being generally used to distinguish research from care (Kass et al. 2013). We apply these characteristics to the DCOG ALL-10 and ALL-11 protocols and consider the implications of our analysis for the moral obligations of physician-investigators in paediatric oncology, with regard to ethical review and informed consent in particular.

5.2 Comparison of DCOG ALL-10 and ALL-11

Based on ethical guidelines for medical research with human beings and scholarly literature, Kass and colleagues have assembled five characteristics of research. These are that research (1) is designed to develop generalisable knowledge and (2) requires systematic investigation. Furthermore, clinical research (3) potentially presents less net clinical benefit and greater overall risk than clinical practice, (4) introduces burdens or risks from activities that are not otherwise part of patients' clinical management, and (5) uses protocols to dictate which therapeutic or diagnostic interventions a patient receives (Kass et al. 2013). We will apply these general research criteria to the two DCOG ALL-protocols in order to explore whether these protocols have research elements. See Fig. 5.1 for an overview of the differences and similarities between the two protocols.

5.2.1 Research Is Designed to Develop Generalisable Knowledge

The first characteristic of research is that research is designed to develop generalisable health knowledge. This characteristic is mainly based on research ethics guidelines such as the *Belmont Report* and the *International Ethical Guidelines for Biomedical Research involving Human Subjects* of the Council for International Organizations of Medical Sciences (CIOMS) (CIOMS 2002; NCPHSBBR 1979).

Similarities between DOCG ALL-10 and ALL-11
- Protocols for the treatment of Acute Lymphoblastic Leukaemia (ALL)
- Both conducted by the Dutch Childhood Oncology Group (DCOG)
- Children aged 1–19 years old
- With newly diagnosed ALL
- Treatment determined by detailed protocols
- Three risk groups based on initial response to therapy (Minimal Residual Disease (MRD) levels)
    - Standard Risk (SR)
    - Medium Risk (MR)
    - High Risk (HR)
- Intensity of treatment based on risk group (i.e., prognosis)
    - Decrease in therapy for SR patients
    - Increase in therapy for MR and HR patients
- Three phases of treatment
    - Induction (Protocols IA, IB and M)
    - Intensification (SR → protocol IV, MR → intensification 1 and 2, HR→ 6 HR blocks or 3 HR blocks and Stem Cell Transplantation)
    - Maintenance
- Use of the same variety of chemotherapeutic agents

Differences between ALL-10 and ALL-11

ALL-10	ALL-11
• Single-arm treatment protocol	• National multicentre open-label randomised clinical trial (Phase III)

• No randomisations	• Two randomisations:
	  – Continuous vs. non-continuous dosage of PEG-Asparaginase
	  – Prophylactic administration of immunoglobulins

• Inclusion from 2004 to 2012, 780 patients included	• Inclusion from 2012 to 2018, 630 patients expected
• E. coli Asparaginase in induction	• PEG-Asparaginase in induction
• Same treatment for patients with Down syndrome	• Different treatment for patients with Down syndrome
• Standard dose of PEG-Asparaginase	• Lowered starting dose and individualised dosage of PEG-Asparaginase based on drug monitoring programme
• Total Body Irradiation for HR patients who receive Stem Cell Transplantation	• No Total Body Irradiation for HR patients who receive Stem Cell Transplantation (chemotherapy instead)

Fig. 5.1 Comparison of the DCOG ALL-10 and ALL-11 protocols

At present, the primary aim of both ALL-10 and ALL-11 is to improve the overall treatment results for children with ALL in terms of Event Free Survival (EFS) compared to the previous DCOG ALL-protocols. The ALL-10 protocol contains several hypotheses about the treatment that is provided to different groups of patients. The protocol states that its aim is to investigate whether these hypotheses will be confirmed. For example, in patients with good prognoses the aim of the study is to investigate whether therapy reduction is feasible without increasing the risk of relapse. To assess whether these improvements have occurred, the outcome of the different patient groups is compared to the historical control groups and international groups of patients from the German BFM (Berlin-Frankfurt-Munster) Study Group. Another part of the protocol is being performed in collaboration with the Australian and New Zealand Children's Cancer Study Group (ANZCCSG), which is said to be necessary in order to obtain sufficient patient numbers to produce statistically significant results. This international collaboration indicates the scientific objective of the ALL-10 protocol.

In addition to the treatment part of the protocols, ALL-10 and ALL-11 encompass several non-therapeutic research studies, to gather data on pharmacokinetics, pharmacodynamics, side effects, etc. As such, these research projects contribute to generalisable knowledge on ALL, drug characteristics and treatment effects. Furthermore, in the ALL-11-protocol a scientific goal is clearly present in the two randomisations, which are aimed at gaining knowledge on dosage of cancer drugs (Asparaginase) and the effects of prophylactic administration of immunoglobulins.

Nonetheless, Dutch paediatric oncologists consider the ALL-10 and ALL-11 protocols as best available treatment, because their treatment regimens are based on the knowledge and experience of the Dutch and international paediatric oncology community at that time. The protocols have implemented the latest insights of the field. Consequently, the objective of producing generalisable knowledge is integrated with the objective of delivering best available treatment for patients.

In sum, although ALL-10 and ALL-11 have been developed to provide state-of-the-art treatment for patients with ALL, they are also designed to produce generalisable knowledge on ALL treatment and the advancement of therapy for children with ALL as a group. Thus, the first defining feature of research applies to both leukaemia protocols.

5.2.2 *Research Requires Systematic Investigation*

The second characteristic of research involves "the systematic collection of data according to a predefined method ... important [for] the production of generalisable knowledge" (Kass et al. 2013, 7). The ALL-10 and ALL-11 protocols are evaluated on a number of measures, such as Event Free Survival (EFS), Disease Free Survival (DFS), Overall Survival (OS), Cumulative Incidence of Relapse (CIR), side effects, and serious adverse events. To facilitate studying these factors, systematic collection of clinical and epidemiological data and storage of collected tissue is necessary. All data from patients are collected in a database, which contains data from all paediatric oncology centres in the Netherlands that treat patients with ALL. These data are systematically evaluated at the end of the running time of the protocols. For this registration and storage, informed consent is obtained from parents and adolescent patients.

In addition, for the ALL-11 protocol a drug monitoring programme was developed. Serum levels of a regularly used cancer drug (Asparaginase) are measured in all patients at pre-established times, in order to determine whether the Asparaginase dosage is appropriate. Low levels decrease the chance of survival, while high levels increase the risk of toxicity. These serum level data are used to effectively manage the care of individual patients while they are simultaneously employed to assess whether such a drug monitoring programme improves outcomes for ALL patients as a group.

Thus, both protocols satisfy the second characteristic of research, since they involve systematic collection and investigation of treatment results, side effects, adverse events, serum levels and other patient data, which are used to contribute to knowledge about leukaemia treatment. However, these data are also used for adapting and improving therapy for individual patients treated according to these protocols.

5.2.3 *Research Potentially Presents Less Net Clinical Benefit and Greater Overall Risk Than Clinical Practice*

In this section we analyse two aspects of ALL-10 and ALL-11 treatment that are of importance when considering the risks and expected benefits. First, therapy in both ALL-10 and ALL-11 is tailored to the risk of relapse. This means that patients who are at high risk of relapse and with a poor prognosis receive more intensive therapy than the ones with better prognoses. Second, irradiation of High Risk patients who receive a stem cell transplant is omitted from the ALL-11 protocol.

5.2.3.1 Tailoring of Therapy

In ALL-10, patients are stratified into three risk groups: Standard Risk (SR), Medium Risk (MR) and High Risk (HR). This classification is primarily based on response to chemotherapy, most importantly the amount of residual leukaemia cells that can be detected with molecular techniques at different times during treatment, the so-called Minimal Residual Disease (MRD). ALL-10 is the first Dutch ALL-protocol to make use of these new molecular techniques and MRD levels to stratify patients. Several studies have shown the clinical relevance of the detection of very low numbers of residual leukaemic cells (Szczepanski et al. 2001). A landmark study by Van Dongen et al. has demonstrated that MRD levels can distinguish "patients with good prognoses from those with poor prognoses, and this helps in decisions whether and how to modify treatment" (van Dongen et al. 1998, 1731). Thus, risk group stratification is used to determine intensity of treatment.

For children in the Standard Risk group the treatment had been reduced compared to previous protocols in order to decrease the burden of the treatment while maintaining survival rates of more than 90%. Lowering the intensity of treatment could have enormous benefits for the patients, in terms of fewer or less severe side effects and a decrease in late effects of treatment when children mature. Improving quality of life for cancer survivors is an important aspect of current anti-cancer therapy (SKION 2010). The ALL-10 treatment strategy for Standard Risk patients turned out to be quite successful: survival rates for this group of patients were very high (>95 %) without additional risk of relapse.

Patients in the Medium Risk group and the High Risk group received a much more intensive chemotherapy regimen than Standard Risk group patients and patients from previous ALL-protocols. This intensification of treatment could mean great advantages in survival (Nachman et al. 1998). However, since cancer drugs are toxic medications, more severe side effects were likely to occur in patients in the MR and HR groups, while it was uncertain whether the goal of increased survival rates would be achieved.

During the course of the ALL-10 protocol it was noticed that toxicity was severe, especially for patients with Down syndrome, leading to a relatively high number of deaths due to side effects. Therefore, a part of the treatment regimen was made less intensive to decrease treatment-related adverse events. These changes are maintained in ALL-11. Although the exact magnitude of the toxicity for patients was unexpected, severe toxicity in patients with Down syndrome had been previously reported. Increased sensitivity of patients with Down syndrome to some chemotherapeutic agents (especially methotrexate) had already been shown in 1987 (Whitlock 2006) and is currently a well confirmed attribute of this group of patients (Peeters and Poon 1987).

Summarising, risk group stratification by Minimal Residual Disease levels and the subsequent intensification of treatment, although based on a variety of international studies, was a novel approach in the Netherlands when implemented at the start of ALL-10. Therefore, the level of the risks of this new approach was mostly unknown and could be expected to be considerable. Hence, the ALL-10 protocol satisfies the third characteristic of research. The treatment regimen of ALL-11 is closely based on ALL-10, which means that during the development of the ALL-11 protocol, interim results on the effectiveness of increasing therapy and appropriate dosage of medications were available. However, data from one study are not sufficient to provide conclusive evidence. Consequently, the risks and uncertainties of benefits of the treatment intensification seem to indicate that the third characteristic of research also applies to ALL-11. To further assess the validity of this conclusion, we describe another aspect of the ALL-11 protocol that involves uncertainty about risks and benefits.

5.2.3.2 Total Body Irradiation

In ALL-10, Total Body Irradiation (TBI) is used to prepare High Risk patients for Stem Cell Transplantation (the conditioning regimen). Due to the risks of several severe side effects, TBI is omitted from ALL-11 and a conditioning regimen of three different chemotherapeutic agents is introduced instead. In order to assess whether omission of TBI is a safe option, the non-inferiority of this conditioning regimen is monitored during the progress of ALL-11.

The ALL-11 protocol reviews several studies investigating the risks and benefits of different drugs compared to TBI. The protocol concludes that a regimen using a busulfan, fludarabin and clofarabin regimen can safely replace the TBI regimen, but data on its efficacy and the long-term side effects are lacking. Moreover, recent data suggest that regimens including TBI might even be preferred over regimens with chemotherapeutic drugs alone. A 2011 review comparing a regimen with TBI and one chemotherapeutic agent to a regimen with two chemotherapeutic agents states that "there is conflicting data on the superiority of one regimen over the other" (Gupta et al. 2011, 17). The review shows that the regimen that includes TBI is favoured over the other regimen. Also, another study of TBI concludes that "conditioning for bone marrow transplantation without radiation is an attractive option, but is not sufficiently effective to completely replace TBI for the most common paediatric indications" (Linsenmeier et al. 2010).

Thus, the studies discussed in the ALL-11 protocol give conflicting answers on the optimal conditioning regimen of HR patients prior to receiving Stem Cell Transplantation. At the moment of implementation of the ALL-11 protocol it was not clear whether TBI could be safely replaced by a chemotherapy conditioning regimen. Omission of TBI in ALL-11 is associated with several uncertainties regarding the risks and benefits, which satisfies the third characteristic of research.

5.2.4 Research Introduces Burdens or Risks from Activities That Are Not Otherwise Part of Patient Care

The ALL-10 and ALL-11 protocols include several research studies that are not part of the treatment of patients, which are reviewed by a Research Ethics Committee, and for which written informed consent is required from parents and, if applicable, patients themselves.

For the additional studies of the ALL-10 protocol, extra blood needs to be drawn, which is generally done during regular blood draws needed for diagnosis and treatment decisions. In addition, patients are requested to collect some buccal tissue using a cotton swab (five times). Furthermore, parents and adolescents are asked to keep a diary during the course of the treatment, to note fever and infection occurrences.

Also, for the ALL-11 research studies, extra blood needs to be drawn, generally during regular blood draw times. In addition, it may be necessary for patients to remain in the hospital 2–4 h longer in order to administer blood compounds. Buccal tissue needs to be collected for one research question.

As we have already explained, the ALL-11 protocol includes two randomisations. The PEG-Asparaginase randomisation does not involve any additional burdens or risks, since no extra interventions have to be performed. The immunoglobulin randomisation involves the collection of extra blood during regular blood draws every 6 weeks. In addition, parents are asked to register on a website to record whether their child had a fever and whether the child had to be admitted to the hospital due to fever.

To summarise, both protocols involve extra research questions, which pose some additional risks and burdens upon patients and parents. The extra time investment or collection of tissue related to answering research questions would otherwise not have been necessary for the treatment of patients. Hence, both ALL-10 and ALL-11 exhibit this fourth characteristic of research.

5.2.5 Research Uses Protocols to Dictate Which Therapeutic or Diagnostic Interventions a Patient Receives

The majority of paediatric oncology treatments are given according to detailed protocols regardless of their categorisation as research or treatment (Verschuur 2004). The same holds for the ALL-protocols. Both ALL-10 and ALL-11 are protocols that describe in detail which medications should be given to patients at which phase of treatment. The treatment laid down in these protocols is based on up-to-date evidence provided by medical scientific research and clinical trials, both on adults and children. This evidence is collectively assessed and discussed extensively within the paediatric oncology community. Therefore, even when no decisive evidence is available at the moment of implementation of new protocols, a variety of sources are employed to determine next steps in treatment in order to provide optimal therapy to current patients and to further increase survival percentages.

Studies have shown that the use of strict and extensive treatment protocols improves the end result of that treatment. In the 1990s Bleyer already recognised the benefits of treatment according to protocols (Bleyer 1997). De Vries and colleagues note that this benefit "would be due to the explicit description of treatment phases and follow-up and to strict guidelines indicating how to deal with side effects and relapses" (de Vries et al. 2011, 7).

Providing treatment according to such detailed protocols does not mean that these protocols are followed blindly. Since all patients are closely monitored, the treating paediatric oncologist can, usually after consulting colleagues, decide to make individual adaptations on the basis of treatment results or side effects. So, although these pre-established protocols in principle determine treatment, patient care can be individualised.

In short, ALL-10 and ALL-11 provide treatment to patients according to well-defined and extensive protocols and thereby satisfy the fifth characteristic of research. However, protocol-controlled treatment is very common in paediatric oncology and has been shown to be beneficial to individual patients. Also, if medically indicated, patients can receive individualised care adapted to their needs. Hence, following strict protocol guidelines simultaneously serves a scientific purpose and the individual treatment needs of patients.

5.2.6 Summary

The two ALL-protocols meet all five characteristics that are generally used to distinguish research from care. Both protocols are designed to develop and to contribute to generalisable knowledge; employ systematic investigation of collected data; have uncertainties with regard to the level of risks and benefits; introduce burdens or risks from activities that are not otherwise part of patient care; and make use of strictly defined protocols to determine treatment. However, three of these characteristics seem compatible with a characteristic of standard treatment as well. First, in addition to the scientific objective, both protocols also involve the goal of providing state-of-the-art treatment for current patients. The treatment regimens of the ALL-protocols are based on national and international scientific studies and consensus within the Dutch paediatric oncology community. Second, patients who receive treatment according to ALL-11 stand to benefit from the systematic data collection, most notably from the Asparaginase drug monitoring programme. Third, treatment according to strict protocols has been shown to improve their results compared to treatment determined by individual physicians and is therefore also beneficial to patients. Thus, the two ALL-protocols do not seem to fall neatly into one of the two categories of research and treatment. Yet, from a traditional bioethical perspective, we have to conclude that both protocols deserve to be categorised as research, mainly due to the relative uncertainty with regard to the level of risks and benefits.

5.3 Discussion

Within the current ethical framework we can only conclude that the correct course of action should have been to regard the ALL-10 protocol as a research protocol with appropriate research ethics review. The reasons are the relatively uncertain level of risks associated with the innovative elements of the treatment regimen and the explicit scientific goal of evaluating this new treatment regimen.

However, if we regard ALL-10 as research, then the question arises, in what respect would research ethics review have improved the protection of patients being treated according to the ALL-10 regimen? Although more stringent regulatory requirements would have been applicable, such as national legislation (WMO 2006) and European regulations (CoE 2005; EP 2001, 2006, 2014), these would not have added measures that had not already been taken. Monitoring of patients was quite extensive, a Data Safety Monitoring Board was installed, and a protocol for reporting serious adverse events (SAEs) and adverse events (AEs) was in place. Determining that the ALL-protocol was a research protocol would not have improved the monitoring. Also, because of the strict monitoring, the treatment regimen of the protocol could be adapted for subgroups of patients in case of multiple SAEs. Hence, even if the protocol had been considered research and had been submitted to the REC concerned, as a research protocol, it would most likely not have improved the protection of patients from harm.

Another way in which categorisation as a research protocol could provide an extra safeguard for patients, is the requirement for an elaborate informed consent process, finalised by signing the informed consent document. Generally, written informed consent is required only for medical research, while presumed or oral consent is acceptable for treatment (Grady 1991). In the case of ALL-10, parents and older patients had already been asked to provide written informed consent for receiving treatment according to the protocol. However, one could argue that the categorisation of a certain activity, that is, as research or standard treatment, could alter the informed consent process, because it has an impact on the mindset of paediatric oncologists: categorising a certain activity as research implies uncertainties, while a standard treatment label implies that the risks and benefits are relatively well known and proportionate. Since a label is never neutral, this might influence the way physicians present the information with respect to a certain protocol. As such, the "standard treatment" label of the ALL-10 protocol from our case could have influenced the way paediatric oncologists presented this protocol to parents and patients, possibly making paediatric oncologists less sensitive to conveying uncertainties. Parents and patients should have been informed of all the relevant aspects of the ALL-10 protocol, including the experimental nature of elements of the treatment regimen, to enable them to provide valid informed consent (AAP 1995). Consequently, if the ALL-10 protocol had been considered a research protocol this might have improved the informed consent process.

Furthermore, if the protocol had been regarded as research, patients and their parents should have been given the choice whether or not to participate. In theory, patients or their parents might have asked for the treatment regimen of a previous protocol. Some parents and adolescents might have favoured more established therapy for which survival rates and side effects had already been evaluated. However, for paediatric oncologists it would be unthinkable to offer an older protocol as well, since they commonly believe it is unethical to withhold a certain treatment from patients if the entire paediatric oncology community regards it as best available treatment. As soon as a new protocol is implemented, the previous protocol is considered outdated. They will always prefer offering the treatment regimen of the new protocol to offering the treatment of a previous protocol. For paediatric oncologists offering something other than the ALL-10 protocol does not amount to a meaningful choice. Rather, it would mean delivering suboptimal care in order to give patients and parents freedom of choice.

With regard to the ALL-11 protocol, we believe that it was not solely its two randomisations that should have led to its categorisation as research. There was also considerable uncertainty about the merits of omission of Total Body Irradiation as a conditioning regimen for patients who have to undergo Stem Cell Transplantation. Although good reasons support omitting TBI, especially the severity of its side effects, data on the comparative risks and benefits of Total Body Irradiation and a chemotherapy regimen are uncertain. Normally, a reasonable option would have been to design a randomisation to compare the different kinds of conditioning regimens. However, due to the relatively low number of HR patients who receive a Stem Cell Transplantation, conducting a randomisation was impossible. Hence, the decision of Dutch paediatric oncologists to leave TBI out of the conditioning regimen is understandable. Still, we argue that decisions to alter a part of a treatment regimen for which the evidence is non-conclusive calls for categorisation as research, since it will demonstrate the relative uncertainty accompanying the decision to leave out TBI.

In the current research ethics paradigm, with its strict distinction between research and care, both ALL-protocols should be regarded as research. However, in the future it remains to be discussed how to review hybrid protocols that integrate a research goal with the objective of simultaneously providing patients with best current treatment. In line with Kass and colleagues we think that a change in both research ethics regulation and oversight of conventional care is needed. We should strive for a research oversight system that is able to do at least two things. First, it should accommodate hybrid protocols and other practices that integrate research with care. This system may call for a different manner of review and may have implications for the informed consent process when research and care turn out to be inseparable. Second, ethically, interventions that are considered standard of care should also be reviewed, due to the absence of available data on safety and effectiveness.

The initial scope of our findings is modest, since we have only discussed two protocols for the treatment of ALL in the Netherlands. However, ALL is the most common form of cancer in children (Pieters 2010; Stiller 2009). Also, combining research and care is standard practice in international paediatric oncology (de Vries et al. 2011), which means that protocols such as those for the treatment of ALL are not unique and our findings are potentially generalisable. Additional case studies could help to determine whether the distinction between research and care in paediatric oncology should be upheld.

5.4 Conclusion

Even though research and treatment are being combined for the benefit of the individual and groups of patients, both ALL-protocols should now be regarded as research protocols since they satisfy five characteristics of research. Yet, in the future it remains to be discussed how to review hybrid protocols that integrate a research goal with the objective of providing individual patients with best current treatment. A change in both research ethics regulation and oversight of conventional care is needed. Further case studies are essential to deepen the moral evaluation of the intertwinement of research and care in paediatric oncology.

References

Ablett, S., and C.R. Pinkerton. 2003. Recruiting children into cancer trials—role of the United Kingdom Children's Cancer Study Group (UKCCSG). *British Journal of Cancer* 88: 1661–1665.

American Academy of Pediatrics (AAP). 1995. Informed consent, parental permission, and assent in pediatric practice. *Pediatrics* 95: 314–317.

Beauchamp, T.L. 2011. Viewpoint: Why our conceptions of research and practice may not serve the best interest of patients and subjects. *Journal of Internal Medicine* 269: 383–387.

Bleyer, W.A. 1997. The U.S. pediatric cancer clinical trials programmes: International implications and the way forward. *European Journal of Cancer* 33: 1439–1447.

Brody, H., and F.G. Miller. 2003. The clinician-investigator. Unavoidable but manageable tension. *Kennedy Institute of Ethics Journal* 13: 329–346.

Conroy, S., C. Newman, and S. Gudka. 2003. Unlicensed and off label drug use in acute lymphoblastic leukaemia and other malignancies in children. *Annals of Oncology* 14: 42–47.

Council for International Organizations of Medical Sciences (CIOMS). 2002. *International ethical guidelines for biomedical research involving human subjects*. Geneva.

Council of Europe (CoE). 2005. *Additional protocol to the convention on human rights and biomedicine, concerning biomedical research*. Strasbourg.

de Vries, M.C., M. Houtlosser, J.M. Wit, D.P. Engberts, D. Bresters, G.J. Kaspers, and E. van Leeuwen. 2011. Ethical issues at the interface of clinical care and research practice in pediatric oncology: A narrative review of parents' and physicians' experiences. *BMC Medical Ethics* 12: 18.

European Parliament (EP). *Directive 2001/20/EC of the European Parliament and of the Council* (Apr 4 2001).

European Parliament (EP). *Regulation (EC) No 1901/2006 of the European Parliament and of the Council* (Dec 12, 2006).

European Parliament (EP). *Regulation (EU) No 536/2014 of the European Parliament and of the Council* (Apr 16, 2014).

Grady, C. 1991. Ethical issues in clinical trials. *Seminars in Oncology Nursing* 7: 288–296.

Gupta, T., S. Kannan, V. Dantkale, and S. Laskar. 2011. Cyclophosphamide plus total body irradiation compared with busulfan plus cyclophosphamide as a conditioning regimen prior to hematopoietic stem cell transplantation in patients with leukemia: A systematic review and meta-analysis. *Hematology/Oncology and Stem Cell Therapy* 4: 17–29.

Joffe, S., and J.C. Weeks. 2002. Views of American oncologists about the purposes of clinical trials. *Journal of the National Cancer Institute* 94: 1847–1853.

Kass, N.E., R.R. Faden, S.N. Goodman, P. Pronovost, S. Tunis, and T.L. Beauchamp. 2013. The research-treatment distinction: A problematic approach for determining which activities should have ethical oversight. *The Hastings Center Report* 43(s1): S4–S15.

Kimland, E., P. Nydert, V. Odlind, Y. Bottiger, and S. Lindemalm. 2012. Paediatric drug use with focus on off-label prescriptions at Swedish hospitals – A nationwide study. *Acta Paediatrica* 101: 772–778.

Krueger, G. 2008. *Hope and suffering children, cancer and the paradox of experimental medicine*. Baltimore: Johns Hopkins University Press.

Largent, E.A., S. Joffe, and F.G. Miller. 2011. Can research and care be ethically integrated? *The Hastings Center Report* 41: 37–46.

Lindell-Osuagwu, L., M.J. Korhonen, S. Saano, M. Helin-Tanninen, T. Naaranlahti, and H. Kokki. 2009. Off-label and unlicensed drug prescribing in three paediatric wards in Finland and review of the international literature. *Journal of Clinical Pharmacy and Therapeutics* 34: 277–287.

Linsenmeier, C., D. Thoennessen, L. Negretti, J.P. Bourquin, T. Streller, U.M. Lutolf, et al. 2010. Total body irradiation (TBI) in pediatric patients. A single-center experience after 30 years of low-dose rate irradiation. *Strahlentherapie und Onkologie* 186: 614–620.

Nachman, J.B., H.N. Sather, M.G. Sensel, M.E. Trigg, J.M. Cherlow, J.N. Lukens, et al. 1998. Augmented post-induction therapy for children with high-risk acute lymphoblastic leukemia and a slow response to initial therapy. *New England Journal of Medicine* 338: 1663–1671.

National Commission for the Protection of Human Subjects of Biomedical and Behavioral Research (NCPHSBBR). 1979. *The Belmont Report. Ethical principles and guidelines for the protection of human subjects*. Department of Health, Education, and Welfare.

Pandolfini, C., and M. Bonati. 2005. A literature review on off-label drug use in children. *European Journal of Pediatrics* 164: 552–558.

Peeters, M., and A. Poon. 1987. Down syndrome and leukemia: Unusual clinical aspects and unexpected methotrexate sensitivity. *European Journal of Pediatrics* 146: 416–422.

Pieters, R. 2010. Acute lymphoblastic leukaemia in children and adolescents: Chance of cure now higher than 80 %. *Nederlands Tijdschrift voor Geneeskunde* 154: A1577.

Stichting Kinderoncologie Nederland (SKION) Richtlijn follow-up na kinderkanker. SKION; 2010.

Stiller C. 2009. *Incidence of childhood leukaemia*. World Health Organization Europe.

Szczepanski, T., A. Orfao, V.H. van der Velden, J.F. San Miguel, and J.J. van Dongen. 2001. Minimal residual disease in leukaemia patients. *The Lancet Oncology* 2: 409–417.

't Jong, G.W., A.G. Vulto, M. de Hoog, K.J. Schimmel, D. Tibboel, and J.N. van den Anker. 2000. Unapproved and off-label use of drugs in a children's hospital. *New England Journal of Medicine* 343: 1125.

van den Berg, H., and N. Tak. 2011. Licensing and labelling of drugs in a paediatric oncology ward. *British Journal of Clinical Pharmacology* 72: 474–481.

van Dongen, J.J., T. Seriu, E.R. Panzer-Grumayer, A. Biondi, M.J. Pongers-Willemse, L. Corral, et al. 1998. Prognostic value of minimal residual disease in acute lymphoblastic leukaemia in childhood. *Lancet* 352: 1731–1738.

van Thiel, G.J.M.W., and J.J.M. van Delden. 2010. Reflective equilibrium as a normative empirical model. *Ethical Perspectives* 17: 183–202.

Verschuur, A.C. 2004. De plaats van klinische trials in de kinderoncologie. *Ned Tijdschr Oncol* 1(1): 30–33.

Weijer, C., and P.B. Miller. 2003. Therapeutic obligation in clinical research. *The Hastings Center Report* 33: 19–28.

Whitlock, J.A. 2006. C. *British Journal of Haematology* 135: 595–602.

WMO Medical Research involving Human Beings Act (Mar 1, 2006).

Chapter 6
What Does the Child's Assent to Research Participation Mean to Parents? Empirical Findings in Paediatric Oncology in Germany

Imme Petersen and Regine Kollek

Abstract National law in Germany requires that, whenever possible, children must provide their assent before participating in clinical research. However, there is still academic debate about many fundamental components of assent in order to address, for example, the age or stage of development respectively, at which children should be asked for assent. Furthermore, only a few studies approach the child's assent to research participation empirically. We present empirical findings from a population-based survey in Germany on parents whose children were first diagnosed with childhood cancer in 2005. The survey's primary objective was to evaluate what the child's assent to research participation means to parents who gave consent on behalf of their minor child. In particular, we wanted to better understand what parents think about the requirement of seeking assent, how to assess the children's competence to give assent and who should be in charge of it. Our empirical findings indicate that parents want to give children a voice in the decision-making regarding research participation. Even though the child's competence to rationally understand the research protocol is primarily discussed in the literature as the most important precondition for a valid assent, the surveyed parents emphasise the child's maturity instead. Given that maturity is regarded as a gradual process, parents want to have a say in assessing it. From this, it follows that parents develop and use a decision-making model that establishes appropriate roles, individual choices and responsibilities for the children, the parents and the physicians.

I. Petersen (✉) • R. Kollek
University of Hamburg, Research Centre for Biotechnology, Society and the Environment, Lottestrasse 55, 22529 Hamburg, Germany
e-mail: imme.petersen@uni-hamburg.de

6.1 Introduction

The survival rates for most childhood cancers have increased tremendously during recent decades. For all childhood cancers combined, the 5-year-survival rate has risen from less than 20 to about 80% in the developed countries during the past 30 years (Howlader et al. 2014; Kaatsch 2010). This improvement has been closely tied to systematic research efforts in paediatric oncology. Due to the fact that childhood cancer is rare and the insight that evidence from cancer research done with adults cannot be transferred directly to children, paediatric oncology has developed a strong research culture (de Vries et al. 2011, 18). One of its features is the widespread enrolment of children who are suffering from cancer in clinical trials. In Germany, for instance, over 90% of all children with cancer under 18 years participate in a clinical trial during treatment and indeed, it is generally considered standard of care to do so.

Most treatments are provided according to national or international protocols that represent the best available treatment according to the current literature, but may also include research components (e.g., randomisation, variable doses of medication) that purportedly result in potential improvements in the treatment. Due to the integration of research and care, paediatric oncology always faces ethical challenges, especially regarding consent procedures for research participation. First, the integration of research and care impedes the parents' and children's ability to differentiate the scientific goals and treatment objectives of a trial (Broome et al. 2001; Chappuy et al. 2008, 2010; Kodish et al. 2004). Second, many treatment and research decisions have to be made under time pressure and emotional distress due to the life-threatening diagnosis (Dermatis and Lesko 1990; Sloper 1996). Third, research protocols can be extremely complex with a considerable textual variety among them (Joffe et al. 2006). Therefore, the consent documents explaining these protocols are often hard to understand for laypersons (Berger et al. 2009).

Before a child is enrolled in a cancer trial, permission must always be obtained from the parent(s) or legal guardian. Many guidelines also require the child's assent, which is defined as "the affirmative agreement to research participation" of children who are capable of providing it (e.g., Code of Federal Regulations 1991; CIOMS 2002; CoE 2005). Assent actively involves the child and obliges doctors to provide information about the proposed research at a level he or she can comprehend and use to make a voluntary choice (De Lourdes et al. 2003, 629). As in many other countries, German law requires that children must provide their assent before participating in research studies whenever possible.

However, there is still debate about fundamental components of assent (Carroll and Gutmann 2011; Unguru et al. 2008). Controversies include the age or stage of development at which investigators should routinely ask children for assent (Martenson and Fägerskiöld 2008; Miller et al. 2004); how much and what information children need (de Vries et al. 2010; Larcher and Hutchinson 2010); methods for assessing both children's understanding of disclosed information and of the assent itself (Hein et al. 2012); who should be involved in the assent process

(McKenna et al. 2010); and, how to resolve disputes between children and their parents (de Vries et al. 2009, 2010).

Despite growing support for empirical studies examining children's understanding of what it means to participate in and agree to research, and their preference for involvement in research (Broome et al. 2001; Chappuy et al. 2008; Susman et al. 1992), only a few studies exist addressing how parents, who usually give consent on behalf of their minor child, assess the child's assent for research participation (Geller et al. 2003). This paper attempts to address this gap by presenting empirical findings from a population-based survey in Germany on parents whose child was first diagnosed with childhood cancer. In order to explore how the parents evaluate children's assent to research participation, we will focus on the following four aspects: (1) What do parents think about the requirement of seeking assent? (2) How should children's competence to give assent be assessed? (3) Who should be in charge of it? (4) Finally, how should a child's refusal to participate in research be dealt with?

6.2 Empirical Approach

We carried out a standardised survey among 1465 parents whose children were first diagnosed with a malignant disease or a central nervous system tumour in 2005. The survey was conducted from March 1 to July 15, 2009, with the help of the German Childhood Cancer Registry (GCCR). As a national population-based registry, it aims to collect data on all cancer cases for children under 15 years (since 2009: under 18 years) in Germany (Kaatsch et al. 1995).[1] With the consent of parents or legal guardians, about 95% of all German children subject to these conditions are registered in the GCCR by name, as reported by the paediatric oncology units.

All registered families with a child under the age of 15 who was first diagnosed with a disease defined in the International Classification of Childhood Cancer (ICCC-3) (Steliarova-Foucher et al. 2005) between January 1 and December 31, 2005 were eligible for inclusion in our survey. As is the rule, most of them were treated in a diagnosis-specific clinical trial, and only a few (5%) were treated off-trial. Children who had died were included, with the exception of children who had died within 6 months before the planned contact. The surveys were approved by the ethics committee of the Medical Association of Hamburg, Germany.

The mail-based survey was conducted in cooperation with the German hospitals that had treated the children in 2005 and had reported the cases to the GCCR. The hospitals were given the opportunity to exclude individual patients from the survey (e.g., due to anticipated emotional distress in the family). In the letter accompanying

---

[1] Based on the Segi WHO world standard, the GCCR uses age-specific incidence rates for children under age 1, ages 1–4, ages 5–9 and ages 10–14. From 2009 onward adolescents aged 15–17 are also included. For further information on registry methods please visit: http://www.kinderkrebsregister.de/dkkr-gb/about-us/overview.html?L=1 (Accessed 15 Feb 2015).

the questionnaire, the parents were assured that the information would be processed anonymously and would be destroyed 5 years after collection. If no response was received by 4–6 weeks, the GCCR sent a single written reminder. If families had moved to an unknown address, the GCCR attempted to trace them through the registry office.

The questionnaire was designed based on a review of existing empirical studies on child's assent and parental consent (e.g., Chappuy et al. 2008; Geller et al. 2003; Miller et al. 2004). In addition to questions regarding the informed consent process of the parents, we asked the parents whether or not their child had given assent to participate in a clinical trial. To assess parents' attitude towards the child's assent, we wanted to know if and, if so, on what terms a child should be asked for assent, who should be in charge of it and how should a child's refusal to participate be dealt with. The questions underwent a content validity assessment with paediatricians at the Department of Paediatric Hematology and Oncology at the University Medical Centre Hamburg-Eppendorf where the questionnaire was piloted with 10 parents of a child less than 15 years who had been diagnosed with a disease defined in the ICCC-3.

The questionnaires were sent to 1494 families. However, some families (n = 29) had moved to an unknown address and could not be traced. Finally, 1465 questionnaires were mailed successfully, of which 807 questionnaires could be evaluated (response rate: 55.1 %). The identifiers were removed from the returned questionnaires, which were subsequently stored electronically only if consent had been given. The questionnaire entailed closed-ended questions with discrete answer variables that were coded into numbers and transcribed into a data matrix. Statistical analysis of the data was conducted using SPSS 16.0. The study results are presented as descriptive statistics. The subgroups were compared using Pearson's chi-squared test. Our survey results regarding parental informed consent are presented elsewhere (Petersen et al. 2013).

6.3 What Does the Child's Assent to Research Participation Mean to Parents? Empirical Findings in Paediatric Oncology in Germany

6.3.1 What Do Parents Think About the Requirement of Seeking Assent?

In order to enrol a child in a clinical trial in a European country, it is obligatory to ask parents or legal guardians to give consent after receiving proper information about the trial, its requirements and risks (see Article 4(c) EU Directive 2001/20/EC). In addition, the German drug law (AMG) requires that the parental consent

Table 6.1 Minor patients' year of birth (median: 2000)

1990	1.6%	(13)
1991	3.5%	(28)
1992	6.1%	(49)
1993	5.9%	(48)
1994	4.7%	(38)
1995	4.0%	(32)
1996	6.2%	(50)
1997	5.6%	(45)
1998	6.2%	(50)
1999	5.7%	(46)
2000	7.4%	(60)
2001	10.4%	(84)
2002	9.2%	(74)
2003	9.0%	(73)
2004	9.3%	(75)
2005	5.2%	(42)
Total	100.0%	(807)

Table 6.2 Did your child give his/her assent to take part in a clinical study?

Yes	16.9%	(136)
No	59.4%	(479)
Don't know	11.0%	(89)
Total	87.2%	(704)
No answer	12.8%	(103)

must correspond to the child's presumed will and that the child's assent must be sought whenever possible (see § 40 s. 4 (3) AMG). Irrespective of the child's age, the attending paediatric oncologist has to inform the child according to its capacity to understand the trial as well as the possible risks and benefits. However, little is known about the current assent practices at clinical sites. We assumed that the investigators primarily approach the parents to make the decision on behalf of their child as many of these children are under the age of 5 years when they are first diagnosed (Kaatsch 2010).

Correspondingly, the median age of the children in our survey was 5 years at the time of first diagnosis in 2005 (see Table 6.1). Of the 704 families who had reported that their child was enrolled in a clinical trial, 19.3% of them stated that their child gave assent to participate in the trial (see Table 6.2). 68.0% of the surveyed parents answered that their child did not give assent and 12.6% were unsure about the child's assent. As anticipated, the amount of given assents increases with age. Thus, 13.9% of the children reported to have given assent were under the age of 6 years (year of birth: 1999–2005), while 86.1% of the children reported to have given assent were 7 years old and older (years of birth: 1990–1998).

6.3.2 How to Assess the Children's Competence to Give Assent?

In legal guidelines and ethical discourses, the term "assent" is generally used as the expression of the child's will to participate in a clinical study. However, it is not entirely clear what the given assent means with regard to rational decision making and voluntary choice and under what conditions investigators should routinely ask children for assent. If, for instance, we expect the child to make a judgement about the risks and benefits of the trial before he or she gives assent, such a capacity may not develop before mid-adolescence. However, if the child simply needs to agree based on his or her own perspective and life-experience (e.g., the pain of having a blood test), a much younger child would be capable of assent (Roth-Cline et al. 2011, 235). Furthermore, it is important to notice that the abilities required for making an informed decision vary with the complexity of the study. Some straightforward studies may be accessible to children younger than 9 years, and some complex studies may be inaccessible even to 14-year-olds (Joffe et al. 2006, 866).

Children are considered to have the assent capacity when they understand the nature of the trial and the potential consequences of it (Field and Behrman 2004; Rossi et al. 2003; Unguru et al. 2008; Weithorn and Scherer 1994). In essence, a child should comprehend why he or she is being asked to participate and what the implications of participation will be. Accordingly, the provision of child-focused information regarding the nature of the trial and what it will involve is required (Roth-Cline and Nelson 2013, 296). Hence, separate information sheets and assent forms are thought to be necessary to provide information in age-appropriate language (e.g., Gross 2010; Larcher and Hutchinson 2010). Furthermore, children asked for participation need the capacity to use the information while making an informed and voluntary decision. According to current knowledge, rational decision-making capacity grows continuously with advances in cognitive development during late childhood through adolescence, in which he or she develops the ability to reason abstractly about hypothetical situations, to reason about multiple alternatives and consequences, to combine multiple variables in complex ways, and to examine information in a systematic manner (DeHart et al. 2004; see also Piaget 1929).

Tara L. Kuther (2003, 346f.) elaborated two problems with regard to child capacity. First, children tend to understand illness in nonspecific ways, as they do not differentiate between the symptoms and causes of illness. Rather, they view illnesses as transmitted magically or caused by moral misbehaviour. The children's understanding of illness evolve during development; in adolescence at the latest, more advanced conceptions of illness appear connecting illness with specific symptoms and diseases. Second, young children in particular tend to view authority figures such as physicians and parents as being authorised and powerful and are therefore likely to comply with their requests. As voluntariness is seen as a capacity that emerges with social and emotional maturity (e.g., Scherer and Reppucci 1988), adolescents are more likely to question parental demands and to decide without being constrained by others.

From this it follows that children need to develop cognitive, emotional and social capacity to give a rational and voluntary assent. All three aspects of this capacity together can be defined as "maturity". Only mature minors meet the conditions to give assent (De Lourdes Levy 2003, 631). Several empirical studies have assessed children's capacity to understand research-related information or their comprehension of actual trials (Abramovitch et al. 1995; Broome and Richards 2003; Dorn et al. 1995; Fernandez 2003; Geller et al. 2003; Ondrusek et al. 1998; Susman et al. 1987, 1992; Tait et al. 2003; Weithorn and Campbell 1982). These studies suggest that children under the age of 9 generally have difficulties understanding research-related information, for instance, research goals and procedures, risks and alternatives. Adolescents aged 14 years and older can reach the level of understanding expected from adults under optimal circumstances. Accordingly, the major and most rapid changes and individual variability in children's capacities occur between the ages of 9 and 14.

In our survey, we wanted to know how German parents assess the child's involvement in the decision-making process. Asked for their opinion regarding the necessity of assent, only very few of the surveyed parents (3.6%) thought that children should generally not be involved in the consent process or that children should always be involved (4.6%). More than 90% stated that the child's assent is necessary if he or she is mature enough to nearly (44.5%) or fully (46.1%) understand the goal and course of research in which they are asked to participate. Confirming the outcome of previous studies, the respondents thought that maturity develops in the course of childhood: only 7.2% of the surveyed parents assumed that children under the age of 9 are mature enough to independently decide whether or not to participate in research, whereas 32.5% of the parents stated they would cede the decision to participate to the child aged between 9 and 12. Approximately the same proportion of parents (37.5%) wanted to assign the consent for research participation to early adolescents (12–16 years) and 17.8% to late adolescents (16–18 years); very few (4.9%) did not want to let the child independently decide until he or she is of full age (see Tables 6.3 and 6.4).

Our results suggest that the assent capacity is thought to differ within same age groups, as the parents associated children's maturity with very different age levels ranging from 9 to 18 years with no clear cut age brackets. Therefore, as a rule, it seems to be necessary to individually assess the child's maturity. This is in line with

Table 6.3 Should children give their assent regarding whether or not they want to participate in a clinical study?

Yes, they should always assent	4.6%	(37)
Yes, when they are mature enough to nearly understand what is going on	44.5%	(359)
Yes, when they are mature enough to fully understand what is going on	46.1%	(372)
No	3.6%	(29)
Don't know	1.2%	(10)
Total	100%	(807)

Table 6.4 When are children on average mature enough to independently decide regarding their study participation?

	Valid percent	
Younger than 9 years	7.2 %	(57)
9–12 years	32.5 %	(257)
12–16 years	37.5 %	(297)
16–18 years	17.8 %	(141)
Older than 18 years	4.9 %	(39)
Total	100.0 %	(791)

the position of developmental psychologists of the post-Piaget era (e.g., Kuther 2003). They assume that no clearly defined developmental stages exist, but that development is an individual process. Some children may be fully involved in the consent process and be able to give informed assent, whereas others will feel more comfortable delegating the decision to their parents. To date, it is not possible to scientifically explore how capacity develops and why individual differences exist. It is therefore recognised that age is, at best, only a proxy for developmental capacity and that experience, maturity and psychological state are key determining factors (Hein et al. 2012, 156).

6.3.3 Who Should Assess Assent Capacity?

In legal guidelines, it remains vague as to how to assess the child's maturity adequately (see, for example, the wording of Art. 6 (2) of the Convention for the Protection of Human Rights and Dignity of the Human Being with regard to the Application of Biology and Medicine 1997). If mentioned at all, age, maturity, and psychological state are considered to be adequate criteria to measure the child's assent capacity (e.g., U.S. Department of Health and Human Services, 45 CFR 46, 1983). The German drug law only requires that the patient must be informed by a physician who has experience in dealing with minors (see § 40 s. 4 (3) AMG). Therefore, the assessment of maturity is usually left to the attending paediatric oncologist.

We wanted to know what the surveyed parents thought of who should be the person in charge of assessing the capacity and maturity of the individual child asked to give assent. The majority of parents wanted the parents to be involved in this process: 26.3 % of the respondents stated that only the parents should decide on their child's maturity. However, the majority thought it would be appropriate to cooperate with the treating physician (63 %) and a few with a psychologist (1.5 %) or with both parties (7 %). In contrast, having physicians, psychologists or both assessing the capacity of the child to assent without the parents is not an option for the respondents (psychologist: 0.9 %; attending physician: 0.6 %; psychologist and physician: 0.8 %).

Our results indicate that German parents believe in the importance of participating in the assessment of their child's assent capacity. However, in the clinical setting, parents experience less involvement in this process than they desire (McKenna et al. 2010). According to research on parents of hospitalised children, physicians normally underestimate the level of involvement and independence that parents wish to exert on the decision-making process and caretaking during a child's illness (Shields et al. 2004). However, nearly one third of the parents surveyed in our study thought that the assessment of their child's assent capacity should be entirely in parental hands. Nearly twice as many wanted to be involved in a shared decision model together with the attending physician. The parents' preference for shared decision-making is in line with the notion of partnership research wherein it is recognised that parental involvement is at best realised by strategies to facilitate parental decision-making processes (Tomlinson et al. 2006).

6.3.4 How to Deal with a Child's Refusal to Participate?

Although children are at the centre of the assent process, parents and physicians have an important influence on the child's decision-making. As the physicians normally evaluate the child's assent capacity, their assessment directly affects children's level of involvement. Recent research has shown that physicians tend to judge a child as competent if the child's decision conforms to the physicians' own ideas of the child's best interest (de Vries et al. 2009, 2010). Obviously, the physicians assessed the capacity of the child rather by the content of the decision than by the process of reasoning in deciding about participation. At the same time, there is empirical evidence that parents influence their child's decision on study participation and that in practice a child's decision is rarely made independently of his or her parents (Bluebond-Langner et al. 2010; Scherer 1991). This likely explains why situations in which children and their parents fundamentally disagree about research participation are rather uncommon (Joffe et al. 2006, 867).

Influencing the extent of a child's involvement in decision-making may avoid conflicts between children and their parents or physicians. However, even if the child is involved, it is not yet clear whether his or her decision carries the same power as his or her parents' and if, from an ethical perspective, the child may veto the parents' decision (Unguru 2011, 201; see also Baylis et al. 1999). In this respect, the German drug law only codifies that any minor's expression of unwillingness to take part in the clinical trial must be respected (§ 40, sec. 4 (3) AMG). However, the chosen wording ("to be respected") does not shed light upon the question as to whether the refusal of a child who lacks legal standing should be as authoritative as the decision of his or her parents.

In our survey, many parents did not want their child to be enrolled in research without any restrictions. Of the surveyed parents, 45 % were willing to respect the

Table 6.5 Should children take part in a clinical study even though they refused?

No, because the child's refusal should be respected	45 %	(363)
Yes, because the cancer has to be treated	29.5 %	(238)
Don't know	15.7 %	(127)
Yes, because children don't understand how important the study is	3.5 %	(28)
Yes, because this will help other children	3.1 %	(25)
Yes, because the parents made the decision	1.5 %	(12)
No, because this would be psychologically stressful	1.0 %	(8)
No, because treatment off-trial is available	0.5 %	(4)
Yes, because the child profits from previous studies	0.2 %	(2)
Total	100 %	(807)

child's refusal irrespective of age; very few parents (1 %) did not want to enrol the child who had refused because otherwise the enrolment would put psychological pressure on the child or because treatment and cure are also possible without study enrolment (0.5 %). However, nearly one third of the respondents (29.5 %) were against this position, stressing that the diagnosed cancer is best treated in a diagnostic-specific clinical trial or simply because the parents have the power to decide (1.5 %). Some parents referred to the absence of maturity (3.5 %) and, finally, very few parents pointed to some kinds of obligations (to help other children: 3.1 %; the child has benefited from previous studies: 0.2 %). Among parents surveyed, 15.7 % were unsure about how to deal with the child's refusal (see Table 6.5).

Our results suggest that German parents hold a strong all-or-nothing position regarding the question of how to deal with the child's refusal to participate in research: nearly half of the respondents were willing to accept the child's refusal, but about one third of surveyed parents could not accept the child's refusal, putting the parents' own ideas of the child's best interest first. To resolve such fundamental disagreement between parents and their child, it seems necessary to arrange procedures for addressing such conflicts. For example, Joffe and his colleagues (2006, 867) suggest appointing an advocate for the child, asking an IRB member or other individual to serve as a neutral consent monitor, and requesting an ethics consultation. The consultant's objectives should be to ensure that the voices of all parties—especially that of the child—are heard and to facilitate agreement between the parents, the child and the physician whenever possible (Aulisio et al. 2000).

6.4 Some Conclusions

Children belong to vulnerable groups in research and are therefore in need of protection. Giving children a role in decision-making by asking for assent is one way of ensuring this safeguard. At the same time, obtaining assent shows respect for a child's dignity and developing personality. Ideally, to respect a child's decision requires an appreciation for what he or she understands and for his or her preferences (Unguru et al. 2008, 217). The current ethical debate focuses on the child's cognitive abilities and his or her capacity to rationally understand the research protocol. Even if these are important aspects of assent, our empirical findings indicate that the surveyed parents tend to put the child's preferences forward, whether they are rational or not. Most of the surveyed parents wanted to give children a voice in the decision-making if and when they are sufficiently mature to nearly or fully understand the goal and course of research in which they are asked to participate. Since, in the view of the survey moiety, the understanding of the research protocol needs to be achieved only approximately, rational understanding is therefore not accepted as an exclusive requirement for assent. This group seems to have a broader understanding of maturity, that is, maturity beyond cognitive capacity; they want to share the decision regarding research participation with the child as early as possible.

There are situations, however, where some of the surveyed parents evaluated the child's protection as being higher than the child's decisional authority. Even if half of the survey participants declared that they were willing to respect the child's refusal irrespective of age, about 30 % of the respondents did not accept the child's refusal to participate in a clinical paediatric oncology trial and put the parents' own ideas of the child's best interest first. From this it follows that assent is often seen contextually, including the child's capacity and the circumstances under which the child's assent can be solicited.

Furthermore, our empirical findings suggest that assent capacity is regarded as increasing with the child's development. However, as the survey participants associated children's maturity with very different age levels ranging from 9 to 18 years, it seems necessary as a rule to individually assess the child's maturity, in order to avoid the mistake of imposing complex research decisions on children who are unable to make them, or inadvertently excluding capable children who want to take part (Joffe et al. 2006, 865). To escape these pitfalls, most of the German parents believe that it is in the child's best interest when they, who know the child best, are involved in the assessment process.

Combining our empirical findings, we conclude that German parents are more than willing to give children a voice in decision-making regarding research participation in paediatric oncology. However, a substantial number of parents are hesitant to grant children a share in decision-making authority. In practice, a child's decision is rarely made independent of his or her parents. It would be a mistake failing to appreciate how most families function and not recognising the interrelated nature of child-parent decision-making. To deal appropriately with individual authority and responsibility, parents appear to need and often use a shared decision-making model that establishes appropriate roles for themselves, their child and the attending physicians. At least in the German setting, there is some variability as to which model the surveyed parents prefer. It therefore seems necessary to put the balance between parents, children and physician in the centre of the discussion about assent.

References

Abramovitch, R., J.L. Freedman, K. Henry, and M. Van Brunschot. 1995. Children's capacity to agree to psychological research: Knowledge of risks and benefits and voluntariness. *Ethics and Behavior* 5(1): 25–48.

Aulisio, M.P., R.M. Arnold, and S.J. Youngner. 2000. Health care ethics consultation: Nature, goals, and competencies. A position paper from the society for health and human values-society for bioethics consultation task force on standards for bioethics consultation. *Annals of Internal Medicine* 133(1): 59–69.

Baylis, F., J. Downie, and N. Kenny. 1999. Children and decision making in health research. *IRB* 21(4): 5–10.

Berger, O., B.H. Grønberg, K. Sand, S. Kaasa, and J.H. Loge. 2009. The length of consent documents in oncology trials is doubled in twenty years. *Annals of Oncology* 20(2): 379–385.

Bluebond-Langner, M., J.B. Belasco, and M. DeMesquita Wander. 2010. "I want to live, until I don't want to live anymore": Involving children with life-threatening and life-shortening illnesses in decision making about care and treatment. *The Nursing Clinics of North America* 45(3): 329–343.

Broome, M.E., and D.J. Richards. 2003. The influence of relationships on children's and adolescents' participation in research. *Nursing Research* 52(3): 191–197.

Broome, M.E., D.J. Richards, and J.M. Hall. 2001. Children in research: The experience of Ill children and adolescents. *Journal of Family Nursing* 7(1): 32–49.

Carroll, T.W., and M.P. Gutmann. 2011. The limits of autonomy: The Belmont report and the history of childhood. *Journal of the History of Medicine and Allied Sciences* 66(1): 82–115.

Chappuy, H., F. Doz, S. Blanche, J.C. Gentet, and J.M. Tréluyer. 2008. Children's views on their involvement in clinical research. *Pediatric Blood & Cancer* 50(5): 1043–1046.

Chappuy, H., A. Baruchel, G. Leverger, C. Oudot, B. Brethon, S. Haouy, et al. 2010. Parental comprehension and satisfaction in informed consent in paediatric clinical trials: A prospective study on childhood leukaemia. *Archives of Disease in Childhood* 95(10): 800–804.

Code of Federal Regulations 116–117 CoFRC. 1991. *Federal policy for the protection of human subjects (Subpart A)*. Washington, DC: United States Department of Health and Human Services.

Council for International Organizations of Medical Science (CIOMS). 2002. *International ethical guidelines for biomedical research involving human subjects*. Geneva.

Council of Europe (CoE). 2005. *Convention on human rights and biomedicine*. Additional protocol concerning biomedical research. Strasbourg.

De Lourdes, Levy M., V. Larcher, and R. Kurz. 2003. Ethics working group of the Confederation of European Specialists in Paediatrics (CESP). Informed consent/assent in children. Statement of the ethics working group of the Confederation of European Specialists in Paediatrics (CESP). *European Journal of Pediatrics* 162(9): 629–633.

de Vries, M.C., D. Bresters, D.P. Engberts, J.M. Wit, and E. van Leeuwen. 2009. Attitudes of physicians and parents towards discussing infertility risks and semen cryopreservation with male adolescents diagnosed with cancer. *Pediatric Blood & Cancer* 53(3): 386–391.

de Vries, M.C., J.M. Wit, D.P. Engberts, G.J. Kaspers, and E. van Leeuwen. 2010. Pediatric oncologists' attitudes towards involving adolescents in decision-making concerning research participation. *Pediatric Blood & Cancer* 55(1): 123–128.

de Vries, M.C., M. Houtlosser, J.M. Wit, D.P. Engberts, D. Bresters, G.J. Kaspers, et al. 2011. Ethical issues at the interface of clinical care and research practice in pediatric oncology: A narrative review of parents' and physicians' experiences. *BMC Medical Ethics* 12: 18.

DeHart, G.B., L.A. Sroufe, and R.G. Cooper. 2004. *Child development: It's nature and course*. London/New York: McGraw-Hill Companies, Inc.

Dermatis, H., and L.M. Lesko. 1990. Psychological distress in parents consenting to child's bone marrow transplantation. *Bone Marrow Transplantation* 6(6): 411–417.

Dorn, L.D., E.J. Susman, and J.C. Fletcher. 1995. Informed consent in children and adolescents: Age, maturation and psychological state. *The Journal of Adolescent Health* 16(3): 185–190.

Fernandez, C.V. 2003. Context in shaping the ability of a child to assent to research. *The American Journal of Bioethics* 3(4): 29–30.

Field, M.J., and R.E. Behrman (eds.). 2004. *Ethical conduct of clinical research involving children.* Washington: National Academic Press.

Geller, G., E.S. Tambor, B.A. Bernhardt, G. Fraser, and L.S. Wissow. 2003. Informed consent for enrolling minors in genetic susceptibility research: A qualitative study of at-risk children's and parents' views about children's role in decision-making. *The Journal of Adolescent Health* 32(4): 260–271.

Gross, T. 2010. Challenges and practicalities of obtaining parental consent and child assent in paediatric trials. *Regulatory Rapporteur* 7(6): 15–18.

Hein, I.M., P.W. Troost, R. Lindeboom, M.C. de Vries, C.M. Zwaan, and R.J. Lindauer. 2012. Assessing children's competence to consent in research by a standardized tool: A validity study. *BMC Pediatrics* 12: 156.

Howlader, N., A.M. Noone, M. Krapcho, J. Garshell, N. Neyman, and S.F. Altekruse, et al., editors. *SEER cancer statistics review, 1975–2010*. National Cancer Institute. 2014. http://seer.cancer.gov/csr/1975_2011. Accessed 21 Feb 2015.

Joffe, S., C.V. Fernandez, R.D. Pentz, D.R. Ungar, N.A. Mathew, C.W. Turner, et al. 2006. Involving children with cancer in decision-making about research participation. *Journal of Pediatrics* 149(6): 862–868.

Kaatsch, P. 2010. Epidemiology of childhood cancer. *Cancer Treatment Reviews* 36(4): 277–285.

Kaatsch, P., G. Haaf, and J. Michaelis. 1995. Childhood malignancies in Germany—methods and results of a nationwide registry. *European Journal of Cancer* 31A(6): 993–999.

Kodish, E., M. Eder, R.B. Noll, K. Ruccione, B. Lange, A. Angiolillo, et al. 2004. Communication of randomization in childhood leukemia trials. *JAMA* 291(4): 470–475.

Kuther, T.L. 2003. Medical decision-making and minors: Issues of consent and assent. *Adolescence* 38(150): 343–358.

Larcher, V., and A. Hutchinson. 2010. How should paediatricians assess Gillick competence? *Archives of Disease in Childhood* 95(4): 307–311.

Martenson, E.K., and A.M. Fägerskiöld. 2008. A review of children's decision-making competence in health care. *Journal of Clinical Nursing* 17(23): 3131–3141.

McKenna, K., J. Collier, M. Hewitt, and H. Blake. 2010. Parental involvement in paediatric cancer treatment decisions. *European Journal of Cancer Care* 19(5): 621–630.

Miller, V.A., D. Drotar, and E. Kodish. 2004. Children's competence for assent and consent: A review of empirical findings. *Ethics and Behavior* 14(3): 255–295.

Ondrusek, N., R. Abramovitch, P. Pencharz, and G. Koren. 1998. Empirical examination of the ability of children to consent to clinical research. *Journal of Medical Ethics* 24(3): 158–165.

Petersen, I., C. Spix, P. Kaatsch, N. Graf, G. Janka, and R. Kollek. 2013. Parental informed consent in pediatric cancer trials: A population-based survey in Germany. *Pediatric Blood & Cancer* 60(3): 446–450.

Piaget, J. 1929. *The child's conception of the world*. London/New York: K. Paul, Trench, Trubner & Co, ltd.

Rossi, W.C., W. Reynolds, and R.M. Nelson. 2003. Child assent and parental permission in pediatric research. *Theoretical Medicine and Bioethics* 24(2): 131–148.

Roth-Cline, M., and R.M. Nelson. 2013. Parental permission and child assent in research on children. *Yale Journal of Biology and Medicine* 86(3): 291–301.

Roth-Cline, M., J. Gerson, P. Bright, C.S. Lee, and R.M. Nelson. 2011. Ethical considerations in conducting pediatric research. *Handbook of Experimental Pharmacology* 205: 219–244.

Scherer, D.G. 1991. The capacities of minors to exercise voluntariness in medical treatment decisions. *Law and Human Behavior* 15(4): 431–449.

Scherer, D.G., and N.D. Reppucci. 1988. Adolescents' capacitiesto provide voluntary informed consent. The effects of parental influence and medical dilemma. *Law and Human Behavior* 12: 123–141.

Shields, L., J. Hunter, and J. Hall. 2004. Parents' and staff's perceptions of parental needs during a child's admission to hospital: An English perspective. *Journal of Child Health Care* 8(1): 9–33.

Sloper, P. 1996. Needs and responses of parents following the diagnosis of childhood cancer. *Child: Care, Health and Development* 22(3): 187–202.

Steliarova-Foucher, E., C. Stiller, B. Lacour, and P. Kaatsch. 2005. International classification of childhood cancer, third edition. *Cancer* 103(7): 1457–1467.

Susman, E.J., L.D. Dorn, and J.C. Fletcher. 1987. Reasoning about illness in ill and healthy children and adolescents: Cognitive and emotional developmental aspects. *Journal of Developmental and Behavioral Pediatrics* 8(5): 266–273.

Susman, E.J., L.D. Dorn, and J.C. Fletcher. 1992. Participation in biomedical research: The consent process as viewed by children, adolescents, young adults, and physicians. *Journal of Pediatrics* 121(4): 547–552.

Tait, A.R., T. Voepel-Lewis, and S. Malviya. 2003. Do they understand? (part II): Assent of children participating in clinical anesthesia and surgery research. *Anesthesiology* 98(3): 609–614.

Tomlinson, D., M. Capra, J. Gammon, J. Volpe, M. Barrera, P.S. Hinds, et al. 2006. Parental decision making in pediatric cancer end-of-life care: Using focus group methodology as a prephase to seek participant design input. *European Journal of Oncology Nursing* 10(3): 198–206.

Unguru, Y. 2011. Making sense of adolescent decision-making: Challenge and reality. *Adolescent Medicine: State of the Art Reviews* 22(2): 195–206.

Unguru, Y., M.J. Coppes, and N. Kamani. 2008. Rethinking pediatric assent: From requirement to ideal. *Pediatric Clinics of North America* 55(1): 211–222.

Weithorn, L.A., and S.B. Campbell. 1982. The competency of children and adolescents to make informed treatment decisions. *Child Development* 53(6): 1589–1598.

Weithorn, L.A., and D.G. Scherer. 1994. Children's involvement in research participation decisions: Psychological consideration. In *Children as research subjects: Science, ethics, and law*, ed. M.A. Grodin and L.H. Glantz, 133–179. New York: Oxford University Press.

Chapter 7
Assent in Paediatric Research and Its Consequences

Jan Piasecki, Marcin Waligora, and Vilius Dranseika

Abstract This article proposes a consequentialist approach to the problem of children's assent in research. To date, one of the main controversies concerning assent has been about the necessary conditions for making a morally significant decision. Some argue that to make a morally significant decision a child has to understand the abstract concept of altruism. Therefore it is crucial to determine at what stage of development this ability arises. Others argue that the crucial condition is to determine when children gain the capacities for making autonomous decisions regarding participation in research. Since these philosophical and psychological controversies are quite persistent, a calculation of the benefits and harms might be essential for implementing a uniform policy. We argue that the benefits of a properly applied policy requiring assent from all capable children is more beneficial than a policy setting a high age threshold for assent. We also suggest that the consequentialist argument depends on empirical premises that might be either supported or proven false by empirical research.

7.1 Background

The requirement to acquire assent from minors participating in research is currently incorporated in many ethical guidelines and regulations concerning research involving children. This requirement should be distinguished from that of informed consent (McGee 2003; Miller and Nelson 2006; Nelson 2007; Nelson and Reynolds 2003). Assent is not just a counterpart of informed consent. The first purpose of assent is to

---

J. Piasecki • M. Waligora (✉)
REMEDY, Research Ethics in Medicine Study Group, Department of Philosophy and Bioethics, Jagiellonian University, Medical College,
Michalowskiego 12, 31-126 Krakow, Poland
e-mail: m.waligora@uj.edu.pl

V. Dranseika
REMEDY, Research Ethics in Medicine Study Group, Department of Philosophy and Bioethics, Jagiellonian University, Medical College,
Michalowskiego 12, 31-126 Krakow, Poland

Department of Logic and History of Philosophy, Vilnius University, Vilnius, Lithuania

protect vulnerable subjects who are not yet able to make their own decisions concerning participation in medical research. Research involving incompetent and non-capable children requires assent combined with parental permission and the prospect of direct benefit. In the case of research without the prospect of direct benefit the additional limitations of risk and peer benefit are set (Piasecki et al. 2015). All these safeguards replace the necessity of obtaining informed consent. The concept of assent should also be distinguished from that of dissent. Dissent can be expressed verbally, or might not be verbalised by all children involved in studies. Therefore, the requirement of respecting a child's dissent does not imply that a child has the capacity to understand his/her situation and research. In contrast, assent may only be obtained from a child who has at least some basic understanding of his/her situation and the procedures that research involves.

At least two different concepts of assent may be distinguished (Giesbertz et al. 2014). On the one hand, assent can be understood as legally non-binding consent (Fisher 2013). Let us call this approach *independent assent*. It can also be understood as the child's engagement by different degrees in the decision-making process, which might be named *involving assent*. The former concept of assent is justified in the principle of autonomy, and the latter refers to respect for developing autonomy. It is considered an important element of acquiring the ability to be autonomous and an important factor supporting communication between the participant and researcher (Giesbertz et al. 2014). A review of regulations and guidelines confirms that the concept of assent is not fully clear and regulations might refer to both independent assent and involving assent. The concept is formulated in the regulations in many ways, and there are also differences in implementation (Kimberly et al. 2006; Kon 2006; Waligora et al 2014). For instance, the US Code of Federal Regulations (46.402) defines assent as "affirmative agreement", which is not merely absence of objection. EU Directive 2001/20/EC as well as forthcoming Regulation (EU) No 536/2014 concerning clinical trials states that a minor should also receive information, according to his or her capacity, about the benefits and risks associated with the study. Directive states that it is also necessary to take into consideration the minor's explicit wish not to participate in research, if the minor is capable of forming an opinion and assessing information concerning his or her participation in clinical trials (Article 4, Par. b, c). In this case the Regulations requires to respect minor's wish (Article 31, Par 1 c). The EU Directive is supplemented with Ethical Considerations for Clinical Trials on Medicinal Products Conducted with the Paediatric Population issued by an ad hoc group for the development of implementing guidelines for Directive 2001/20/EC. This document advises that whenever appropriate, a child should take part in the process of informed consent with her or his parents. The ad hoc group proposes three different solutions for small children, children who are able to give their assent, and adolescents, whose confidentiality should be protected, especially in the case of research on socially sensitive issues. The UNESCO Universal Declaration on Bioethics and Human Rights does not mention the word "assent", but speaks of involvement in the process of decision-making (Article 7, Par. a). The Convention on Human Rights and Biomedicine requires that potential incompetent participants be informed of all their rights and safeguards

(Article 29, Par. 2, point iii) and states that a minor's opinion should be considered according to his or her age and maturity. The Declaration of Helsinki states that a physician must seek the assent of a potential subject who is deemed incapable of giving consent (Article 29). The lack of cohesiveness of the regulations and guidelines is based on the lack of one definition of assent—definitions are rarely provided, and suggestions as to how exactly this requirement should be realised are absent. Therefore, much room is left for interpretation and implementation of the requirement of assent (Baines 2011; Kimberly et al. 2006; Kon 2006).

One of the main controversies concerning assent is determining capacities that allow one to make morally significant decisions. On the one hand, it can be argued that a child is able to make a morally significant choice when s/he understands the abstract purpose of research and concept of altruism (Wendler 2006; Wendler and Shah 2003). Hence, there is a mature-autonomy model of obtaining assent only from children who are able to make such choices, and for the most part, it is argued, this ability is gained at the age of 14 (Wendler 2006; Wendler and Shah 2003). On the other hand, taking a developing autonomy approach, one can argue that some other aspects of the decision-making process are important, and not just a fully autonomous child's choices should be respected. Developing autonomy, according to this position, is also deserving of our respect, and therefore it is important to involve a child in the decision-making process (Baylis and Downie 2003; Nelson and Reynolds 2003). Involving assent can be obtained from much younger children, for example at the age of seven. We do not argue that this controversy is irrelevant or that it impossible to resolve this problem. We suggest that there are other, consequentialist arguments supporting development of the autonomy policy.

7.2 How Much Autonomy Is Needed to Give Assent?

One can ask these questions: Which competences must a subject have in order to be able to make morally significant decisions? What does one have to know and understand about research to make a reasonable and autonomous decision concerning participation? Some argue that a subject can make a morally significant decision only if s/he understands the proper purpose of the medical research (Wendler 2006; Wendler and Shah 2003). The purpose of research is to produce generalisable knowledge that might help future patients. Its main objective is therefore helping other people. One can make an autonomous decision about research, when the object understands this abstract concept of altruism. He or she does not have to be motivated by altruistic feelings, but has to understand the purpose of the action. It can be argued that one is not merely used only when one understands and approves the proper purpose of use of one's body. One is merely used, otherwise.

But there is a doubt about whether understanding of the abstract concept of altruism is a necessary condition for taking a morally significant action (Baylis and Downie 2003). The argument is as follows: there are many morally significant reasons for taking part in research. One can participate in research because of self-interests or in order to develop one's personality. For instance, a subject can decide to participate in research in order to get free cinema tickets. This can be a rational and subjectively valuable reason for taking part in research.

A proponent of Wendler and Shah might argue that Baylis and Downie miss the point of the argument. They do not realise that a person who participates in order to get free tickets and without the knowledge of the purpose of participation is merely being used. But a proponent of Baylis and Downie might argue that such knowledge does not matter. A participant might have a good reason and just ignore other aspects of participation, at least up to the moment when s/he approves experienced burdens. Moreover, it can be argued that the proponent of Wendler and Shah misses the point, confusing assent and informed consent. A child does not have to know all the aspects of research, but has a right to know some aspects, concrete and tangible aspects, those experienced during the medical procedures. This controversy is quite difficult to resolve and has a conceptual affinity with the controversy over the moral status of entities. What features of natural beings are fundamental to their moral status (Warren 2005)? Is it sufficient to be alive? Or is it necessary to have consciousness? Sometimes, instead of directly resolving these kinds of questions, we refer to the consequences of applying a different solution.

Another approach to determining children's capacity to assent is to detect the moment when a child acquires the general capacities necessary for making morally significant decisions. Some argue that these capacities are mainly developed at the age of 14 (Wendler 2006; Wendler and Shah 2003). Others suggest that the capacities necessary for making a morally significant decision concerning participation in medical research are acquired as early as 11 (Tait et al. 2003a, b). There are some who argue that children with chronic diseases have a more profound understanding of their situation and are capable of empathic acts at an earlier age (Alderson 1992, 2007). The problem with the age of acquiring capacities for morally significant decisions is not merely an empirical one of detecting these capacities in children and therefore a lack of sensitive diagnostic tests. It is difficult to determine an age threshold, because we do not know what exactly has to be detected. Do we want to identify when children are able to understand meaningful elements of disclosure, that is, risks, benefits and medical procedures (Tait et al. 2003a, b)? Or perhaps we want children to understand the abstract concept of altruism (Wendler 2006; Wendler and Shah 2003). Moreover, it may even be argued that we can have significant communication with preterm babies (Alderson et al. 2005). We can therefore say that the problem of age threshold is rather secondary to the problem of morally binding capacities. Proponents of mature autonomy argue in favour of independent assent, setting the age threshold high, because they ascribe moral significance to fully autonomous decisions. Adherents of developing autonomy support involving assent, because for them sometimes even very rudimentary abilities to communicate pain and distress have a moral significance.

7.3 Three Policies, Three Different Consequences

We may ask what the consequences of the three different policies on the issue might be. Firstly, we can imagine the possible consequences of a policy setting a high age threshold for assent. Let us call this a *high-age policy*. Secondly, the consequences of a *low-age* threshold *policy* may be considered. We can then decide which policy gives a better account of the benefits and harms, or we can modify and combine these two policies to maximise the benefits and harms that they can bring. This allows us to formulate a third policy proposal. Moreover, we think that this discussion might be a point of departure for an empirical test of different policies. Such a test would support or prove wrong the suggested solutions on the basis of their consequences.

A high-age policy was proposed by Wendler and Shah (Wendler and Shah 2003). All children's dissent is respected, but only 14-year-olds can give their independent assent for research. Although, according to Wendler and Shah, under this requirement children do not have to understand the research and they may only simply react to however the research affects them (Wendler 2008), the effectiveness of the policy is not so clear. There is a concern that children display a tendency to obey and submit to adults. Without informing children of their rights, they might not know that they can refuse participation, and may be afraid of refusing, and contain their feelings and distress. Another concern is that the research personnel might not be sufficiently skilled and willing to detect and recognise signs of dissent. Therefore, many children who are enrolled and have not been informed that they can express their dissent would bridle, crying or screaming, but still experience strong distress and, perhaps, psychological harm. This might undermine their trust in their parents and physicians, and medical institutions. Children might lose trust in their parents and not perceive them as their protectors. For instance, before launching a trial of growth hormone for short stature, there were some hints that children might have perceived themselves as victims of painful and useless medical procedures and felt anger towards their parents and the research staff (Rotnem et al. 1979; Tauer 1994). As some argue, a lack of discussion may lead to many tensions in families (Sibley et al. 2012). This kind of policy can also hinder the development of children's autonomy. They might learn to be submissive, and also might think that they do not have the right to express their feelings and emotions. Another potential unwanted consequence of this policy might be the lesson that children take away. If we do not ask children for assent, we cannot expect and demand that they ask for assent and respect someone's autonomy (Diekema 2003). Therefore, we can imagine (and also probably measure this doing appropriate empirical research), that there will be some number of children who will be harmed by the implementation of a high-age policy. Nevertheless, this policy brings benefits. A high-age policy will guarantee faster development of medicines for children. Firstly, researchers would not be engaged in difficult and sometimes problematic conversations with children, and can enrol all children whose parents grant their parental permission. This is the merit of shortening the recruitment procedure. Secondly, there are probably a significant number of children who, despite having some abilities to assent, do not make a really well-thought-out decision. Asked to participate in research, they express a wish to play computer games or watch "Barney" (Joffe 2003; Wendler 2008). In their case, the research would probably not be harmful. We ask them for assent, but in fact they are unable to make a significant moral decision. Their reasonless behaviour nevertheless has morally significant consequences for the development of science. Therefore, if we do not have to ask for assent from children who are not able to make a significant moral decision, we can enroll the correct number of children in research more rapidly.

The second policy requires *independent assent* from all children who are able to verbally communicate with researchers. There is no threshold age of assent, and researchers are obliged to assess the capacity of each child. If the child is able to express his/her preferences with regard to participation in research, they have to obtain his/her assent. The consequences of this particular policy might be that the process of recruitment will be significantly obstructed. Many children might not really understand the nature of research, and would make irrational choices not grounded in facts. For example, a child might fear lab coats. Because of the obstruction of enrolment, development of research will be slower. Of course, it would be an exaggeration to claim that all research involving children would be thwarted. There are always some obedient children who would follow the wishes of their parents, and a number of children who assent to participate. However, this policy will also bring some benefits. A significant number of children for whom the research might be harmful would not ultimately be harmed. They will drop out either at the beginning or much more easily during the process. We can even assume that the number of children benefiting would be approximately similar to the number of children who would be harmed, in the case of the first policy considered. Moreover, the list of types of harm might also include harm caused by the policy itself. Some argue that asking for assent might be harmful for some children. It might cause tension within the family (Baines 2011) or might be harmful for a child in the sense that a child may prefer not to be involved and informed about his/her illness (Kon 2006).

We can now consider what to do in order to avoid the negative consequences of the first policy. The first solution is to introduce a requirement of independent assent for all capable children in a low-age sense. But we also know the consequences of the *independent assent* requirement in the case of the low-age policy. Unwanted consequences of the low-age policy might be avoided by adopting a high-age policy. A consequentialist approach would require empirical testing of these two policies. But there might be another, middle way that could also be tested empirically. The moderate policy should have the virtue of identifying all children who would be harmed by participation in the research, due to their susceptibility to pain and stress, who would be able to refuse to participate if asked. Furthermore, the search policy should also allow for obtaining assent from children who are not susceptible and would not be harmed by participation, but are not sufficiently mature to fully consider all aspects of their participation. Another virtue of such a policy would be giving them a chance to rely on their parents, if necessary. We suggest that for all children with developing autonomy such a policy employs the concept of involving assent. This policy requires an attempt to involve all capable children in the decision-making process. This solution does not exclude establishing a certain age-threshold, but this issue is beyond the scope of this article and we discuss it elsewhere (Waligora et al 2014). The process of obtaining involving assent should be appropriate to the emotional and cognitive abilities of a child. In this process the child's preferences should be respected and a child should be informed about all aspects of the study that might directly affect him/her. This means that children would usually consult and discuss their decision with their parents. We can therefore expect that the properly adapted policy of assent might facilitate contact between parents and their children. Proper communication between children and parents would prevent having too many dropouts, because children who are supported by their parents would accept some level of pain and discomfort and not experience it as being stressful and especially harmful. This requirement protects children from psychological or other harm; children feel better and safer when they know what is going to happen, and gain trust in the medical profession and their parents, while the proper procedure of assent, instead of creating tension within the family, might facilitate understanding (Sibley et al. 2012). It seems that assent that is properly adjusted, personally and developmentally, has the merit of maximising the benefits of research. This kind of policy protects fragile and sensitive children from being involved in research. It is also beneficial for the children who are involved in research and it does not obstruct the process of recruitment, which consequently does not impede the growth of generalisable knowledge.

7.4 Conclusion

We argue that resolving the controversy over children's assent does not have to rely solely on the concept of autonomy. There is no sufficient conceptual tool to resolve the problem of determinants of a morally significant choice. Some might argue that only a completely autonomous decision based on a full understanding may have moral significance, while others insist that partially autonomous decisions should also be respected. To date, we do not have a commonly accepted conceptual tool to resolve this problem. Therefore, we can try to check the consequences of two different policies with regard to children's assent. The high-age policy allows 14-year-olds to assent and younger children to dissent. We argue that such a policy might cause harm to many children involved in research against their will, but on the other hand it has the merit of speeding up the research process. The low-age policy, which requires assent from all capable children, slows down development of science, but it protects possible participants from different kinds of harm. We conclude that the policy that allows all capable children to assent but at the same time permits a contribution from their parents and respects children's personal preferences and developmental abilities, both protects children from harm and does not significantly slow down the process of scientific progress.

Acknowledgements We would like to thank Ben Koschalka for linguistic editing. This project was funded by the National Science Centre, Poland, DEC-2011/03/D/HS1/01695.

References

Alderson, P. 1992. In the genes or in the stars? Children's competence to consent. *Journal of Medical Ethics* 18(3): 119–124.
Alderson, P. 2007. Competent children? Minors' consent to health care treatment and research. *Social Science & Medicine* 65(11): 2272–2283.
Alderson, P., J. Hawthorne, and M. Killen. 2005. The participation rights of premature babies. *International Journal of Children's Rights* 13: 31–50.
Baines, P. 2011. Assent for children's participation in research is incoherent and wrong. *Archives of Disease in Childhood* 96(10): 960–962.
Baylis, F., and J. Downie. 2003. The limits of altruism and arbitrary age limits. *American Journal of Bioethics* 3(4): 19–21.
Diekema, D.S. 2003. Taking children seriously: What's so important about assent? *American Journal of Bioethics* 3(4): 25–26.
Fisher, H. 2013. Assent to participate in healthcare research. *Current Allergy & Clinical Immunology* 26(3): 145–150.
Giesbertz, N.A., A.L. Bredenoord, and J.J. van Delden. 2014. Clarifying assent in pediatric research. *European Journal of Human Genetics* 22(2): 266–269.
Joffe, S. 2003. Rethink "affirmative agreement", but abandon "assent". *American Journal of Bioethics* 3(4): 9–11.
Kimberly, M.B., K.S. Hoehn, C. Feudtner, R.M. Nelson, and M. Schreiner. 2006. Variation in standards of research compensation and child assent practices: A comparison of 69 institutional review board-approved informed permission and assent forms for 3 multicenter pediatric clinical trials. *Pediatrics* 117(5): 1706–1711.
Kon, A.A. 2006. Assent in pediatric research. *Pediatrics* 117(5): 1806–1810.
McGee, E.M. 2003. Altruism, children, and nonbeneficial research. *American Journal of Bioethics* 3(4): 21–23.
Miller, V.A., and R.M. Nelson. 2006. A developmental approach to child assent for nontherapeutic research. *Journal of Pediatrics* 149(Suppl 1): S25–S30.
Nelson, R.M. 2007. Minimal risk, yet again. *Journal of Pediatrics* 150(6): 570–572.
Nelson, R.M., and W.W. Reynolds. 2003. We should reject passive resignation in favor of requiring the assent of younger children for participation in nonbeneficial research. *American Journal of Bioethics* 3(4): 11–13.
Piasecki, J., M. Waligora, and V. Dranseika. 2015. Non-beneficial pediatric research: Individual and social interests. *Medicine, Health Care, and Philosophy* 18(1): 103–112.
Rotnem, D., D.J. Cohen, R. Hintz, and M. Genel. 1979. Psychological sequelae of relative "Treatment Failure" for children receiving human growth hormone replacement. *Journal of the American Academy of Child Psychiatry* 18(3): 505–520.
Sibley, A., M. Sheehan, and A.J. Pollard. 2012. Assent is not consent. *Journal of Medical Ethics* 38(1): 3.
Tait, A.R., T. Voepel-Lewis, and S. Malviya. 2003a. Do they understand? (part I): Parental consent for children participating in clinical anesthesia and surgery research. *Anesthesiology* 98(3): 603–608.

Tait, A.R., T. Voepel-Lewis, and S. Malviya. 2003b. Do they understand? (part II): Assent of children participating in clinical anesthesia and surgery research. *Anesthesiology* 98(3): 609–614.

Tauer, C.A. 1994. The NIH, trials of growth hormone for short stature. *IRB* 16(3): 1–9.

Waligora, M., V. Dranseika, and J. Piasecki. 2014. Child's assent in research: Age threshold or personalisation? *BMC Medical Ethics* 15: 44.

Warren, M.A. 2005. *Moral status. Obligations to persons and other living things (issues in biomedical ethics)*. Oxford/New York: Oxford University Press.

Wendler, D. 2006. Assent in paediatric research: Theoretical and practical considerations. *Journal of Medical Ethics* 32(4): 229–234.

Wendler, D. 2008. The assent requirement in pediatric research. In *The Oxford textbook of clinical research ethics*, ed. E.J. Emanuel, A. Robert, C. Grady, R.K. Lie, F.G. Miller, and D. Wendler, 661–669. Oxford/New York: Oxford University Press.

Wendler, D., and S. Shah. 2003. Should children decide whether they are enrolled in nonbeneficial research? *American Journal of Bioethics* 3(4): 1–7.

Chapter 8
Ethical Principles in Phase IV Studies

Rosemarie D.L.C. Bernabe

Abstract Phase IV post-marketing studies on a pharmaceutical product have been increasing in number and presumably in importance recently. This growing number of phase IV studies has led to a greater need to examine the applicable ethics at this stage. Building on our previous work on ethics in phase IV studies, we propose that the following ethical principles are indispensable to implementing ethics in phase IV: (A) When discussing the possibility of waiving informed consent (IC), it is necessary to consider such discussions within the sphere of human rights. (B) The fact that there are a variety of phase IV studies is ethically significant. (C) Study type differences warrant different ethical treatment with respect to issues of IC eligibility for waiver, the manner of obtaining IC, and the content of the IC form. (D) The ethical evaluation of phase IV should assume a therapeutic orientation. (E) The weighing of risks and benefits is a shared responsibility between research ethics committees and sponsors/investigators. (F) In balancing risks and benefits, assuming a therapeutic orientation and applying expected utility theory, the following constraints are in order: the benefit utility table must include the direct benefits; the weight of direct benefits cannot be negligible in comparison with other benefits; and the risk table must include both risks due to study participation and to experimental intervention.

8.1 Introduction

Phase IV studies refer to "all studies (other than routine surveillance) performed after drug approval and related to the approved indication" such as "drug-drug interaction studies, dose-response or safety studies and studies designed to support use under the approved indication" (ICH 1997) as well as studies to obtain health economic data. These studies are usually "larger, less technically complicated than pre-registration studies, have fewer inclusion/exclusion criteria and are more likely to include subjective or qualitative end points" (Johnson-Pratt 2007).

R.D.L.C. Bernabe (✉)
University Medical Center Utrecht, Julius Center for Health Sciences and Primary Care, Huispost Str. 6.131, 85500, 3508GA Utrecht, The Netherlands
e-mail: r_bernabe@yahoo.com

© Springer International Publishing Switzerland 2016
D. Strech, M. Mertz (eds.), *Ethics and Governance of Biomedical Research*, Research Ethics Forum 4, DOI 10.1007/978-3-319-28731-7_8

Phase IV studies have traditionally received less attention than those of other phases. For some time, phase IV trials were almost equated to studies that were not as rigorous as the other phases, and may even have been covert marketing ploys to promote a new drug. These were termed "seeding trials" in the past. However, things have changed. The number of phase IV studies has increased recently (Brower 2007). At ClinicalTrials.gov, the number of yearly registered phase IV studies since 2005 has consistently been greater than 1539, a dramatic rise from the less than 73 phase IV trials registered every year from 2000 to 2004.[1]

The increased number of phase IV studies deserves more attention, even in ethics. The Institute of Medicine's *Ethical and Scientific Issues in Studying the Safety of Approved Drugs* dispenses the same message on the need for a research ethics that is applied in the post marketing stage (IOM 2012). In this chapter, building on our previous articles on ethics in phase IV studies, we endeavour to provide a preliminary elaboration of principles that we think are applicable in phase IV.

Because the principles we propose derive from our previous ethical deliberations, the reader will not find a discussion of the various debates on ethical issues in phase IV here. We wish to direct the reader who may be interested in these debates to our earlier articles. Also, the phase IV issues we have examined are those that fall within what van Thiel and van Delden called the "protection paradigm" (van Thiel and van Delden 2008), that is, ethical questions that are nuanced only by the very character of phase IV and not by the various socio-economic contexts. Of course, we are by no means saying that distributive justice issues in this phase are of less importance; distributive justice issues are equally as urgent and important as the problems within the protection paradigm. However, given that foundational ethical issues conceptually precede contextualised issues, we thought it necessary and helpful to limit ourselves to the issues within the protection paradigm. Within this paradigm, we further limited ourselves to issues of informed consent (IC), risk-benefit assessment, and the therapeutic orientation of phase IV.

8.1.1 Informed Consent

Because IC is widely present in the literature, and to contextualise IC in phase IV, in our previous articles (Bernabe et al. 2011a, b; van der Baan et al. 2013), we considered the question of the necessity of IC and the possibility of waiving it in three different phase IV situations. First, we looked at the necessity of IC in phase IV non-interventional studies (Bernabe et al. 2011b). Contrary to the arguments presented in the literature that IC may not be necessary in such studies (Buckley et al. 2007; Al-Shahi et al. 2005; Tu et al. 2004), we demonstrated that IC remains the standard for such studies even though the manner of obtaining consent and the

---

[1] On December 17, 2012 we conducted a search on ClinicalTrials.gov of all clinical trials that were registered from the year 2000 to the year 2011 with the following results (in chronological order, starting from the year 2000): 29, 32, 70, 51, 72, 1939, 1540, 1684, 2093, 1850, 1838, 1794.

content of the IC form differ from those of the earlier phases (Bernabe et al. 2011b). One difference between IC in earlier trials (i.e., phases I–III) and IC in phase IV non-interventional studies is the acceptability of both opt-in and opt-out procedures for the latter studies. Also, given the presence of a substantive justification, the waiving of IC in the latter may be ethically acceptable in "exceptional" circumstances (CIOMS 2008).[2]

Next, we looked at an interesting area in phase IV studies where the only intervention is the randomisation of the participants to the trial drug and the comparator/s while the other procedures remain "standard clinical practice" (Bernabe et al. 2011a). In this latter study, we clarified that "randomisation" and "observational" are strange bedfellows; or better yet, that putting them together is a contradiction and hence such randomised studies should be classified as interventional instead. After this initial clarification, we again asked the question: Is IC necessary? We concluded that in these studies, the waiving of IC is generally not ethically acceptable, though in cases of minimal risk, an opt-out procedure may be ethically defensible.

Lastly, we looked at the case of obtaining the IC of psychiatric patients for the purpose of biobanking their blood for both clinical use and pharmacogenetic research (van der Baan et al. 2013). In this latter study, we concluded that by considering the risks and the decisional competence of these patients, neither the waiving of IC nor an opt-out procedure is ethically justifiable. Instead, an enhanced opt-in procedure is necessary.

8.1.2 Risk/Benefit Assessment

One of the tasks of an ethics committee, when evaluating any proposed study, is to assess its risks and benefits. We aimed to provide guidance on this task specifically for phase IV studies, but to do so, we needed to take a few steps back. First, we raised the issue of the need for clarity and reasonableness in the balancing of risks and benefits (Bernabe et al. 2009). Next, we looked at the two prominent risk/benefit assessment methods (the procedure-level approaches, i.e., the net risk test and the component analysis approach) in research ethics and concluded that the main difficulty of the methods is conflation of the various risk-benefit tasks[3]; we also

---

[2] In the International Ethical Guidelines for Epidemiological Research, CIOMS uses the term "exceptional" to mean that the waiving of IC, though the standard in epidemiological studies, may at times be ethically justifiable. It then goes on to provide circumstances in epidemiological studies when the waiving of IC may be raised as a possibility: "the use of personally non-identifiable materials; the use of personally identifiable materials with special justification; studies performed within the scope of regulatory authority; studies using health-related registries that are authorized under national regulations; and cluster-randomized trials" (CIOMS 2008).

[3] In decision theory and risk studies, risk-benefit assessment is in fact composed of the following activities in chronological order: risk-benefit analysis, risk-benefit evaluation, risk treatment, and decision making (Aven 2008; Vose 2008).

demonstrated that research ethics would gain from incorporating decision theory methods in the various tasks of weighing risks and benefits (Bernabe et al. 2012a). Lastly, we focused on the risk/benefit evaluation task of research ethics committees and demonstrated that expected utility theory,[4] and in particular, Multiattribute Utility Theory,[5] may aid in making the weighing task more explicit and less intuitive (Bernabe et al. 2012b). Only after establishing these did we feel we had a sufficient basis to provide some guidance on the assessment of risks and benefits in phase IV. We shall return to this point later in this chapter.

8.1.3 Therapeutic Orientation in Phase IV Studies

Lastly, we dealt with the therapeutic orientation in phase IV studies. By therapeutic orientation, we refer to the mindset or inclination whereby research is seen in terms of therapeutic morality (Miller and Rosenstein 2003). In discussing this issue, first we needed to determine the status quo. Hence, given the assumptions that phase IV trials should typically aim at informing a clinical decision and that the value of a phase IV trial hinges on its clinical relevance, we looked at the current state of phase IV non-inferiority trials (Bernabe et al. 2013). In this latter study, we demonstrated that although half of the post-authorisation non-inferiority trials we studied reported additional benefit claims, these claims were seldom supported by sufficient data and formal testing to establish statistical significance. After demonstrating that the design of post-authorisation non-inferiority studies should be improved in terms of ensuring and providing evidence for its clinical value, we explored the question of the fiduciary obligation of the physician researcher in phase IV interventional studies (Bernabe et al 2014). In the latter article, we argued that since phase IV trials are by nature and purpose closer to practice that the other phases of drug development, physician-researchers are primarily physicians and secondarily researchers whose fiduciary obligation to their patient-participants remains, even though some aspects of this obligation may have been waived.

8.2 Ethical Principles in Phase IV

In our previous studies, we have dealt with the nuances surrounding phase IV and how ethics ought to deal with these nuances. Even though our study has been limited to the issues within the protection paradigm, we can still mention a few principles (A to F in the following) when implementing ethics in phase IV.

---

[4] Expected utility theory is a decision analysis tool where probabilities are somewhat known. It works as follows: "to each alternative is assigned a weighted average of its utility values under different states of nature, and the probabilities of these states are used as weights" (Hansson 2005).

[5] Multiattribute Utility Theory is a variant of the expected utility theory where personal values are made explicit by assigning numerical weights to them (Baron 2008).

(A) *When discussing issues of IC waiver, it is ethically necessary to consider such discussions within the sphere of human rights*

In Bernabe et al. 2011a, b and van der Baan et al. 2013, we argued that the waiving of IC is a relevant issue in phase IV, especially in non-interventional studies. The waiving of IC in observational studies such as epidemiological studies, while not universally applicable, is circumstantially justified by reasons such as the following: the exposure of the participant to no more than minimal risk; the procedures to be used customarily do not require IC outside the research context; or, when IC poses a threat to the participant's confidentiality (4).

In Bernabe et al. 2011a, we provided two ethical arguments that are relevant to the question and the justification for waiving IC: the generic right argument and the harm argument. The justifications we just mentioned clearly touch on most of the aspects of the harm argument; the concept of minimal risk, for example, avoids the voluntary imposition of appreciable risks to participants' interests. However, these justifications do not explicitly deal with the generic right argument, that is, if we may recall, that all human beings as agents have generic rights. By generic rights we refer to rights that "agents need, irrespective of what their purposes might be, in order to be able to act at all or in order to be able to act with general chances of success" (Beyleveld and Brownsword 2007). Even the argument that waiver may be justified when IC poses a threat to confidentiality is not exactly phrased as a violation of a basic human right; the term "threat" may easily be construed as referring to exposing participants to risk, and hence, an argument that may be simplistically viewed as protection from harm. In what follows, we shall briefly show that the generic rights argument is indispensable if such waiver is only to be invoked in "exceptional situations".

According to the CIOMS International Ethical Guidelines for Epidemiological Studies, the "waiver of individual IC is to be regarded as exceptional, and must in all cases be approved by an ethical review committee" (CIOMS 2008). If this is true, it means that deliberation on exceptionality and the weighing of factors must ensue before such a waiver is approved; a mere checklist approach is insufficient. Such a checklist approach is all the more insufficient if the list only contains the three justifications we mentioned above. These three justifications do not address exceptionality. At most, they add to the factors to be taken into account when weighing whether the waiver is justifiable. However, and more crucially, these factors do not state what is at stake, i.e., they do not speak to the very reason why such a waiver ought to be exceptional in the first place. The generic right argument provides us with this reason: such waivers ought to be granted only in exceptional situations because that which is waived is the generic right of human beings to confidentiality and self-determination in terms of the use of one's (medical) records. Precisely because the issue is the waiving of human rights, a substantive justification must convincingly show that in this particular study waiver is ethically acceptable, for example, and that in this circumstance public health (or other concerns) weighs more heavily than these rights. Hence, though the generic right argument does not provide us with the tools to weigh such matters, it places the issue of waiver within the proper context.

(B) *The fact that there are a variety of phase IV trials is ethically significant*

The range of methodological types of phase IV studies is appreciably wide: from epidemiological studies that use a database, observational studies with subject enrolment, large simple trials with randomisation, to classical randomised controlled trials (RCTs). The variety of purposes of phase IV studies accounts for the methodological variety in this phase. According to the Wiley Encyclopedia of Clinical Trials, the various purposes of phase IV studies are the following: seeding; identification of rare but serious events; identification of long-term side effects; further dose investigation; exploration of further indications for authorised drugs; interaction studies; comparison with other drugs; quality of life and health economics studies (Day 2008). While some of these purposes may be debatable (such as that of seeding trials since the value of these trials is scientifically questionable, or the exploration of further indications since such trials may be categorised as phase IIIB trials), it is still without doubt that the variety of purposes would require various types of trial methodologies. For example, the identification of rare but serious events may best be accomplished via large simple trials or through an observational study that utilises a large database; the identification of long-term side effects seemingly could best be achieved through observational studies (whether through a database or with subject enrolment); further dose investigation would probably require a randomised controlled trial (RCT); interaction studies may utilise observational studies and/or RCTs; drug comparison, whether non-inferiority or superiority, could best be achieved using an RCT; and quality of life and health economics studies customarily use observational studies and/or large simple trials.

This variety in phase IV trials is ethically significant because it accounts for differences in terms of (1) the demands on patient-participants and (2) the extent of compromise in the fiduciary relationship between the physician and the patient.

In terms of the scale of demands on patient participation, depending on the type of phase IV study, the demands on patients differ. An observational study with patient enrolment may require only interviews and/or answering of questionnaires; a large simple trial may require some sort of randomisation while all other interventions remain clinical and therapeutic; and an RCT would require participants to be assessed for eligibility, randomised, undergo a series of tests, and report for follow-ups. These differing demands also denote differing levels of exposure to risk/discomfort.

Regarding the degree of compromise in the fiduciary relationship, in Bernabe et al. 2014 we discussed the fact that physicians who involve themselves in phase IV research remain fiduciaries to their patients; however, certain responsibilities may be waived. Depending on the protocol, the responsibilities of physicians that must be waived for the sake of the trial differ: an observational study ideally should not affect the physicians' fiduciary obligations to the patient-participants, while an RCT would require certain obligations to be waived, such as the obligation to choose the best therapy based on patients' individualised conditions.

Hence, the various types of phase IV studies expose patient-participants to differing degrees of risk/discomfort and differing degrees of compromised fiduciary obligation. Since the differing degrees of risk and compromise are by-products of

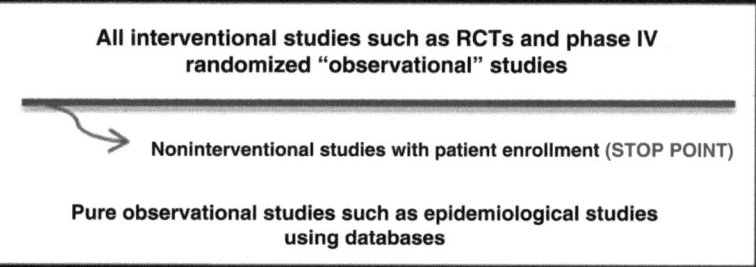

Fig. 8.1 Stop point above which IC in a phase IV study may not be waived

the variety of types of phase IV trials, we can say with confidence that the fact that there are a variety of phase IV trials is an ethically significant one.

(C) *The differences in type warrant different ethical treatment with respect to the issues of IC eligibility for waiver, the manner of attaining IC, and the content of the IC form*

The IC's Eligibility for Waiver In Bernabe et al. 2011a, b and van der Baan et al. 2013, we argued that the acceptability of a substantive justification on the option of waiving the *prima facie* right to consent is limited to phase IV non-interventional drug studies. Once some sort of intervention or complication is present, such as that in the case of phase IV randomised "observational" drug studies (Bernabe et al. 2011a), in pharmacogenetic research using non-anonymous data from psychiatric biobanks (van der Baan et al. 2013), in large simple trials, and of course, in RCTs, then the waiving of IC becomes ethically unjustifiable. We can imagine non-interventional studies with patient enrolment as the stop point, and below this stop point, there would be observational studies such as epidemiological studies that use databases (Fig. 8.1). Above this stop point, IC may not be waived.

Possibility of an Opt-Out System Even though IC may not be waived above the stop point, this does not mean that IC may only be attained through an opt-in system. Depending on aspects such as decisional competence and minimal risk, an opt-out system may be ethically justifiable to accommodate research concerns such as bias and recruitment barriers. In trials where decisional competence may be questionable, such as in pharmacogenetic research with psychiatric biobanks, interventional trials in children, and others, an opt-out system may be ethically unjustifiable. The same can be said about trials with more than minimal risks, such as RCTs.

Difference in the Amount and Kind of Information in the IC Form Article 26 of The Declaration of Helsinki lists the information that ought to be relayed to the patient-participants when seeking their IC:

> In medical research involving human subjects capable of giving informed consent, each potential subject must be adequately informed of the aims, methods, sources of funding, any possible conflicts of interest, institutional affiliations of the researcher, the anticipated

benefits and potential risks of the study and the discomfort it may entail, post-study provisions and any other relevant aspects of the study. The potential subject must be informed of the right to refuse to participate in the study or to withdraw consent to participate at any time without reprisal […]. (WMA 2013).

As stated above, we argued earlier that the amount and kind of information that is needed in securing the IC of patient-participants in phase IV non-interventional studies differs from other types of trials (Bernabe et al. 2011b). For example, in non-interventional studies, the understanding of risks and benefits ought not to be of great concern, at least relative to interventional studies, and this nuance is accounted for by the very nature of non-interventional studies: phase IV non-interventional studies ideally should not have additional risks. Hence, the IC form for such trials would not devote as much space to "risks/discomforts" compared to interventional studies. In this sense, the nature of a phase IV trial dictates the amount and kind of information present in the IC form.

Hence, trial type variability in phase IV accounts for ethically significant nuances in issues such as IC eligibility for waiver, the possibility of an opt-out system, and the differences in the amount and kind of information that ought to be present in the IC form.

(D) *The ethical evaluation of a phase IV study should assume a therapeutic orientation*

In Bernabe et al. 2014, we have shown that by nature and purpose, phase IV trials are closer to practice than other phases. Precisely because the therapeutic orientation of phase IV stems from its very essence, any ethical reflection on phase IV is incomplete without this assumption. This therapeutic orientation is the assumption behind our conclusion on the strength of the fiduciary obligation of physicians in phase IV (Bernabe et al. 2014); it is the same assumption behind the claim that a phase IV study, including non-inferiority trials, ought to aim for clinical significance (Bernabe et al. 2013).

(E) *The weighing of risks and benefits is not the sole task of the research ethics committee; rather, it is the shared responsibility of the sponsor/investigator along with the research ethics committee*

In any study, risks and benefits must be assessed to ensure that risks do not outweigh benefits, or simply, that risks are acceptable in relation to the objectives of the study and the study's (possible) benefits. The task of weighing risks and benefits has traditionally been vested in research ethics committees. In Bernabe et al. 2012a, we agree with this; indeed, it is necessary for research ethics committees to evaluate risks and benefits to ensure that patient-participants are not exposed to unnecessary and unjustified risks and to verify the scientific/social validity of a study. However, it is unfair to assign the entire task of this assessment to research ethics committees. We argued that the weighing of risks and benefits ought to be carried out cooperatively by the sponsor and the research ethics committee: risk/benefit analysis is the task of the sponsor; risk/benefit evaluation is the task of both the sponsor and the research ethics committee; risk treatment is the task of the research ethics committee; and decision making is also the task of the research ethics committee (Bernabe et al. 2012a).

(F) *In the balancing of risks and benefits in phase IV, assuming a therapeutic orientation and applying the expected utility theory (specifically Multiattribute Utility Theory), some ethical constraints are in order*

In all the various risk-benefit tasks, decision studies methods may be helpful in making these tasks "robust, transparent, and coherent" (Bernabe et al. 2012b). This is a conclusion that comports with the findings of the European Medicines Agency Benefit-Risk Methodology Project (EMA 2012). In Bernabe et al. 2012b, we have shown that Multiattribute Utility Theory, when applied to the evaluation of benefits and risks, is capable of accomplishing this aim. When applied specifically in phase IV, it is necessary to presuppose a therapeutic orientation to understand the nuances and constraints that ought to be present in the evaluation of risks and benefits. Earlier in the introduction, we mentioned providing some guidance in the assessment of risks and benefits in phase IV. Guidance here comes in the form of nuances and constraints in phase IV risk/benefits assessment, which include the following:

1. The benefit utility table must necessarily include (potential) "direct benefits." By direct benefits, we mean the benefits that patient-participants receive from the experimental intervention (King and Churchill 2008). By including direct benefits, research ethics committees (or the sponsor) are compelled *to account for* the therapeutic value of the intervention for the patient-participants.
2. The weight of "direct benefits" cannot be negligible in comparison to the weight of other benefits, such as "benefits to society". Admittedly, what qualifies as negligible would subjectively depend on the evaluators; nevertheless, some cases are obvious: on a scale of 1–10, if "benefits to society" weighs 8, the weight of "direct benefits" cannot be 1.
3. The risk table must necessarily include the categories, "risks due to study participation" and "risks due to the experimental intervention" since these risks directly affect the therapeutic value of the trial to the patient-participants.
4. In both categories (i.e., "risks due to study participation" and "risks due to the experimental intervention"), "burdens" and "inconveniences" (whether together or separately) must be considered.

8.3 Conclusion

When implementing ethics in phase IV, we proposed that the following are principles that must be upheld:

- When discussing the possibility of waiving IC, it is necessary to consider such discussions within the sphere of human rights.
- The fact that there are a variety of phase IV studies is an ethically significant one.

- Differences in types of studies warrant different ethical treatment on the issues of IC eligibility for waiver, the manner of obtaining IC, and the content of the IC form.
- The ethical evaluation of phase IV should assume a therapeutic orientation.
- The weighing of risks and benefits is a shared responsibility between research ethics committees and sponsors/investigators.
- In balancing risks and benefits, assuming a therapeutic orientation and employing standard utility theory, the following constraints are, in order: the benefit utility table must include the direct benefits; the weight of direct benefits cannot be negligible in comparison with other benefits; and the risk table must include risks due to study participation and those due to experimental intervention.

8.4 Further Research

This is an initial inquiry into some aspects of research ethics that are relevant in the post marketing stage. Because of its exploratory and preliminary nature, there is much to be done if research ethics is to keep pace with the increasing number of phase IV studies. For example, there is a need to elaborate on the following: how decision theory may aid in the other tasks of weighing risks-benefits, including risk-benefit analysis and risk treatment; how therapeutic orientation may affect phase IV methodology and design; the different degrees of therapeutic orientation in the various phase IV studies; how the formulation of IC ought/ought not to be affected by the therapeutic orientation/fiduciary obligation; the assorted ways that fiduciary obligation is and may be compromised in the various phase IV studies.

References

Al-Shahi, R., C. Vousden, and C. Warlow. 2005. Bias from requiring explicit consent from all participants in observational research: Prospective, population based study. *BMJ* 331: 942.

Aven, T. 2008. *Risk analysis: Assessing uncertainties beyond expected values and probabilities.* Chichester: Wiley.

Baron, J. 2008. *Thinking and deciding*, 4th ed. Cambridge: Cambridge University Press.

Bernabe, R.D., G.J. van Thiel, J.A. Raaijmakers, and J.J. van Delden. 2009. The need to explicate the ethical evaluation tools to avoid ethical inflation. *The American Journal of Bioethics* 9(11): 56–58.

Bernabe, R.D., G.J. van Thiel, J.A. Raaijmakers, and J.J. van Delden. 2011a. Is informed consent necessary for randomized phase IV "observational" drug studies? *Drug Discovery Today* 16(17–18): 751–754.

Bernabe, R.D., G.J. van Thiel, J.A. Raaijmakers, and J.J. van Delden. 2011b. Informed consent and phase IV non-interventional drug research. *Current Medical Research & Opinion* 27(3): 513–518.

Bernabe, R.D., G.J. van Thiel, J.A. Raaijmakers, and J.J. van Delden. 2012a. The risk-benefit task of research ethics committees: An evaluation of current approaches and the need to incorporate decision studies methods. *BMC Medical Ethics* 13: 6.

Bernabe, R.D., G.J. van Thiel, J.A. Raaijmakers, and J.J. van Delden. 2012b. Decision theory and the evaluation of risks and benefits of clinical trials. *Drug Discovery Today* 17(23–24): 1263–1269.

Bernabe, R.D., G. Wangge, M.J. Knol, O.H. Klungel, J.J. van Delden, A. de Boer, et al. 2013. Phase IV non-inferiority trials and additional claims of benefit. *BMC Medical Research Methodology* 13: 70.

Bernabe, R.D., G.J. van Thiel, J.A. Raaijmakers, and J.J. van Delden. 2014. The fiduciary obligation of the physician-researcher in phase IV trials. *BMC Medical Ethics* 15: 11.

Beyleveld, D., and R. Brownsword. 2007. *Consent in the law*. Oxford: Hart.

Brower, A. 2007. Phase 4 research grows despite lack of FDA oversight. *Biotechnology Healthcare* 4(5): 16–22.

Buckley, B., A.W. Murphy, M. Byrne, and L. Glynn. 2007. Selection bias resulting from the requirement for prior consent in observational research: A community cohort of people with ischaemic heart disease. *Heart* 93(9): 1116–1120.

Council for International Organizations of Medical Sciences (CIOMS). 2008. *International ethical guidelines for epidemiological studies*. Council for International Organizations of Medical Sciences: Geneva.

Day, S. 2008. Phase IV trials. In *Wiley encyclopedia of clinical trials*, ed. R. Dagostino, L. Sullivan, and J. Massaro, 446–453. Hoboken: Wiley.

European Medicines Agency (EMA). 2012. *Benefit-risk methodology project*. European Medicines Agency: London.

Hansson, S.O. 2005. *Decision theory: A brief introduction*. Stockholm: Royal Institute of Technology.

Institute of Medicine (IOM). 2012. *Ethical and scientific issues in studying the safety of approved drugs*. Washington, DC: National Academies Press.

International Conference on Harmonisation (ICH). 1997. *General considerations for clinical trials E8*. International Conference on Harmonisation: Geneva.

Johnson-Pratt, L.R. 2007. Phase IV drug development: Post-marketing studies. In *Principles and practice of pharmaceutical medicine*, 2nd ed, ed. L.D. Edwards, A.J. Fletcher, A.W. Fox, and P.D. Stonier, 119–120. Chichester: Wiley.

King, N.M., and L.R. Churchill. 2008. Assessing and comparing potential benefits and risks of harm. In *The Oxford textbook of clinical research ethics*, ed. E. Emanuel, C. Grady, R.A. Crouch, R.A. Lie, F.G. Miller, and D. Wendler, 514–526. New York: Oxford University Press.

Miller, F.G., and D.L. Rosenstein. 2003. The therapeutic orientation to clinical trials. *New England Journal of Medicine* 348(14): 1383–1386.

Tu, J.V., D.J. Willison, F.L. Silver, J. Fang, J.A. Richards, A. Laupacis, et al. 2004. Impracticability of informed consent in the Registry of the Canadian Stroke Network. *New England Journal of Medicine* 350(14): 1414–1421.

van der Baan, F.H., R.D. Bernabe, A.L. Bredenoord, J.G. Gregoor, G. Meynen, M.J. Knol, et al. 2013. Consent in psychiatric biobanks for pharmacogenetic research. *International Journal of Neuropsychopharmacology* 16(3): 677–682.

van Thiel, G.J., and J.J. van Delden. 2008. Phase IV research: Innovation in need of ethics. *Journal of Medical Ethics* 34(6): 415–416.

Vose, D. 2008. *Risk analysis: A quantitative guide*, 3rd ed. Chichester: Wiley.

World Medical Association (WMA). 2013. *Declaration of Helsinki: Ethical principles for medical research involving human subjects*. 64th WMA General Assembly. Fortaleza, Brazil. http://www.wma.net/en/30publications/10policies/b3. Accessed 15 Feb 2015.

Chapter 9
Fate of Clinical Research Studies After Ethical Approval—Follow-Up of Study Protocols Until Publication

Anette Blümle, Joerg J. Meerpohl, Martin Schumacher, and Erik von Elm

Abstract Many clinical studies are ultimately not fully published in peer-reviewed journals. We assembled a cohort of clinical studies approved 2000–2002 by the Research Ethics Committee of the University of Freiburg, Germany. Published full articles were searched in electronic databases and investigators contacted. Data on study characteristics were extracted from protocols and corresponding publications. We characterized the cohort, quantified its publication outcome and compared protocols and publications for selected aspects. Of 917 approved studies, 807 were started and 110 were not. Of the started studies, 576 (71%) were completed according to protocol, 128 (16%) discontinued and 42 (5%) are still ongoing; for 61 (8%) there was no information about their course. We identified 782 full publications corresponding to 419 of the 807 initiated studies; the publication proportion was 52%. Study design was not significantly associated with subsequent publication. Multicentre status, international collaboration, large sample size and commercial or non-commercial funding were positively associated with subsequent publication. Commercial funding was mentioned in 203 (48%) protocols and in 205 (49%) of the publications. In most published studies (339; 81%) this information corresponded between protocol and publication. Most studies were published in English (367; 88%); some in German (25; 6%) or both languages (27; 6%). Half of the clinical research conducted at a large German university medical centre remains unpublished; future research is built on an incomplete database. Research resources are likely wasted as neither health care professionals nor patients nor policy makers can use the results when making decisions.

Originally published as Blümle A, Meerpohl JJ, Schumacher M, von Elm E. Fate of Clinical Research Studies after Ethical Approval—Follow-Up of Study Protocols until Publication. PLoS ONE. 2014;9(2):e87184. doi:10.1371/journal.pone.0087184.

A. Blümle (✉) • J.J. Meerpohl • M. Schumacher
Institute for Medical Biometry and Statistics, Center for Medical Biometry and Medical Informatics, German Cochrane Center, University Medical Center,
Berliner Allee 29, 79110 Freiburg, Germany
e-mail: bluemle@cochrane.de

E. von Elm
Cochrane Switzerland, Institute of Social and Preventive Medicine (IUMSP), Lausanne University Hospital, Lausanne, Switzerland

9.1 Introduction

Patients and health professionals should be able to consider and appraise all the evidence available from medical research in order to make informed decisions about health issues. Such evidence on effectiveness and potential harm of health care interventions comes from interventional and observational studies, published in original articles and well-conducted systematic reviews summarizing primary studies. It has long been known that only a part of all clinical studies ultimately reaches the stage of full publication in peer-reviewed journals (Rosenthal 1979). Publication or non-publication of studies is influenced by factors such as the nature and direction of their results (Easterbrook et al. 1991; Decullier et al. 2005; Dickersin et al. 1992; Stern and Simes 1997). The prevailing underreporting is wasteful and can result in biased estimates of treatment effect or harm (Chalmers and Glasziou 2009).

Prospective trial registration has become an important measure to reduce underreporting by revealing studies that remained unpublished and hidden for the public. While Switzerland makes prospective registration of all human research studies mandatory from 2014 on http://www.kofam.ch/en, this is still not the case in most jurisdictions including the European Union and the USA. Publication outcome is not only influenced by the direction of study results but also by characteristics such as study design and size, funding source or the presence of an international collaboration (von Elm et al. 2008). Main reasons for non-publication are lack of time or low priority, results not deemed important enough and journal rejection (Song et al. 2010).

Consequently, only a particular share of the body of evidence is available to users of research data including other researchers, health professionals and patients. It is given undue prominence in the literature. This can lead to treatment recommendations that are at best inappropriate and at worst dangerous (Dickersin and Chalmers 2010). Selective publication has been deemed unethical, also from a normative point of view (Strech 2012).

Submission to a research ethics committee (REC) or a funding agency is the earliest stage at which a planned study is documented in detail. We set out to assemble an unselected cohort of clinical studies that were approved by the REC of the University of Freiburg/Germany (Albert-Ludwigs-Universität). We aimed to characterize the clinical research being conducted, quantify its publication outcome and compare study protocols and corresponding publications for selected aspects.

9.2 Materials and Methods

9.2.1 Cohort of Study Protocols

We were granted access to the REC's files, which included the protocols of human research studies submitted for ethical approval, amendments, correspondence and other ancillary documents. A first analysis based on 299 protocols of studies of all designs approved in 2000 was published earlier (Blümle et al. 2008). For the present analysis, we completed the cohort of study protocols by adding those approved during the years 2001–2002. The definitive analysis is thus based on the study protocols approved during the three consecutive years 2000–2002. We chose this time period because it was both accessible in the REC's archives and long enough to allow for completion of the included studies. If a study protocol described two or more sub-studies, we regarded each as a separate study.

9.2.2 Data Collection and Definitions

We used a standardised data extraction form (MS Access 2010™) to collect data on study characteristics from the study protocols, amendments (if any), the REC's application forms, and correspondence including study design, sample size, type of funding, single-/multicentre status, leading study centre and domestic/international study status. If conflicting information was found, we recorded the information of the most recent document in our database. If the information was not reported in any of the documents, we classified it as "unclear". Data were extracted by one investigator. If the investigator in charge could not decide on how to extract data (e.g. when classifying study design), the issue was discussed with a second investigator to reach a consensus. All database entries were cross-checked by a second investigator.

We classified studies according to their design using an algorithm established earlier (Blümle et al. 2008). The categories were as follows: randomised controlled trials, non-randomised intervention studies, diagnostic studies, observational studies (incl. cohort, case-control, cross-sectional studies), uncontrolled studies, or laboratory studies (i.e. using human tissue or blood e.g. for genetic research). Funding sources were classified as commercial or non-commercial and information extracted separately. Commercial funding was defined as any direct financial support or provision of material (e.g. of the study drug) by a private for-profit company. We further extracted whether a private company was involved in the planning, management or data analysis of the study. We assumed such involvement if the study protocol was written by its staff or if one of the authors was affiliated with the company. Noncommercial funding was defined as financial or other support by governmental funding agencies, public or private foundations (unless clearly linked to a private company) or research funds of hospitals or academic institutions. We further classified studies as international or domestic. If at least one centre outside Germany participated in recruitment of participants, the study was considered international, otherwise domestic. We extracted the planned overall number of participants to be recruited (study size); if the protocol indicated a range of values we used the smallest value. Information on current study status was collected from correspondence with the applicants or other documents available to the REC.

9.2.3 Identification of Corresponding Publications

We systematically searched the following electronic databases and platforms: Medline (platform Ovid, database Ovid MedlineR+ Daily Update), Web of Science, Google Scholar, Current Contents Medizin including content by the publishers Hogrefe, Karger, Kluwer, Springer and Thieme (combined searches on the Medpilot platform www.medpilot.de) and the University's publication registry (Forschungsdatenbank Freiburg, http://forschdb.verwaltung.uni-freiburg.de/forschung). For randomised controlled trials, we also searched the Cochrane Central Register of Controlled Trials (issues 2/2010–4/2011), which contains records of controlled trials from Medline (quarterly updated), Embase (annually updated) and those identified by manual searches of journals that are not indexed in electronic literature databases (Higgins and Green 2011). A new search strategy was established for each study protocol including keywords from the protocol, such as experimental drug, study name or acronym, studied health condition or names of applicants. We used variants of search terms (e.g. synonyms) and additional search terms (e.g. trade names of drugs or devices) where appropriate. The search strategies were manually adapted to the specific syntax of each literature database. Searches for the protocols of the year 2000 were conducted between July 2011 and January 2012 and included an update of the earlier search conducted in 2006 (Blümle et al. 2008). For the protocols of the years 2001 and 2002, the searches were conducted between August 2009 and January 2010. We retrieved the full text of potentially eligible publications and set up an electronic library of pdf-documents linked to our MS Access 2010™ database. If we came across additional eligible references by other sources (e.g. reference lists of identified articles), we included them. Disagreements on eligibility were resolved by discussion and consensus among the authors. Only articles that contained at least some information on the study's objectives, methods and results and were published in a scientific journal were considered full publications. Review articles and published conference abstracts were excluded. Full reports of preliminary results published before completion of recruitment or data collection as planned were counted as full publications. Retrieved articles were read in full by one investigator. Key elements of study design and methods, but also study acronyms and names of authors, were used as criteria to decide whether the publication was considered matching a study protocol. Any uncertainties were discussed in regular group meetings.

In order to complement the electronic searches, we surveyed the investigators applying to the REC by writing personalised letters. In an appended questionnaire, we asked them for verification of the already identified publications and for references of additional publications we may have missed. We also asked whether the project (a) had been completed as planned (according to the protocol), (b) had been discontinued entirely or at the local study site, or (c) is still ongoing with or without continued recruitment or data collection. The letters and questionnaires were sent out in February 2010 and reminder letters in May 2010. Undeliverable letters were sent out again if the investigators' new address could be determined.

Based on the information from the survey, we checked and updated our publication database by deleting wrongly attributed references and adding any new. We also considered information on the current project status from other sources such as correspondence between the REC and investigators and information from publications. If the information from the survey did not match with what was reported in the publication and could not be clarified otherwise, we used the information from the publication. If we found a corresponding publication by our electronic searches, but received no response in the survey, we used the publication to determine the study's status.

9.2.4 Data Analyses

We used queries in MS Access 2010 and tabulation in Microsoft Excel 2010 to obtain standard descriptive statistics. We calculated the proportion of published study protocols (i.e. the proportion of studies that had been started at the local study site and resulted in at least one corresponding full publication), as well as its binomial 95 % confidence interval. We used Pearson's χ^2 test to examine associations between study characteristics and publication proportion and calculated McNemar odds ratios for disclosure of funding information in pairs of protocols and publications of commercially and non-commercially funded trials (Rothman et al. 2008, 287–288). All comparisons were pre-planned. A p-value of 0.05 was used as threshold for statistical significance. For agreement of funding information between protocols and publications, we calculated Cohen's kappa values with 95 % confidence intervals (Altman 1991).

9.3 Results

Between 2000 and 2002 the REC of the University of Freiburg approved 981 study protocols containing information on 990 individual studies (Fig. 9.1). Seven protocols comprised two substudies and one comprised three sub-studies; we counted each substudy separately. We excluded 73 studies because they were either duplicate submissions from several participating centres or the study was rejected, retracted or an extension of a previous study. Our final dataset comprised 917 approved studies.

9.3.1 Characteristics of Included Studies

Almost half of the submitted studies were randomised controlled trials, which was the most frequent study design (408 studies, 45 %). Of those, most were of parallel design (364 studies, 89 %) with two treatment arms (269 studies) or three or more

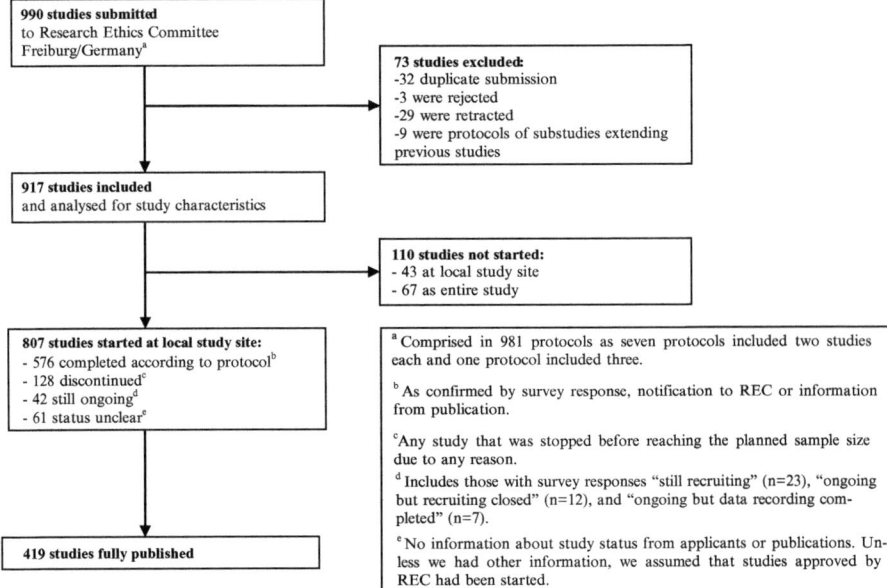

Fig. 9.1 Flowchart of study protocols approved between 2000 and 2002 by the research ethics committee of the University of Freiburg/Germany with number and study status of included studies

treatment arms (95 studies). Twenty-eight studies (7%) had a cross-over design and 16 (4%) another variant design, such as factorial or intra-individual comparison. The second most frequent study design were uncontrolled studies (186 studies, 20%), such as case series or uncontrolled phase I/II studies, followed by laboratory studies using human tissue or blood (138 studies, 15%), nonrandomised intervention studies (72 studies, 8%), cross-sectional studies (42 studies, 5%), diagnostic studies (41 studies, 4%), comparative cohort studies (23 studies, 2%), case-control studies (6 studies, 1%), and one health services research study (0.1%) (Table 9.1). The planned sample size was stated in 878 studies (96%) and ranged from 3 to 9300 participants (median, 120).

The planned duration of enrolment was specified for 382 (42%) studies and ranged from less than 1–120 months (median, 12 months). 383 studies (42%) planned recruitment in a single centre and 534 (58%) in multiple centres. Of the multi-centre studies, 310 (58%) included an international collaboration and 221 (41%) a collaboration with other centres in Germany. Eighty-three multicentre studies (15%) were led by the local investigators and 448 (84%) by other study centres in Germany or abroad. For three studies, the collaboration status and leadership role remained unclear. Of the 221 domestic studies, 49 (22%) were led by the investigators in Freiburg and 171 (77%) by another study centre in Germany. For two studies (one international, one domestic), the leading centre was not determined at the time of REC submission and for one international study there was no intention to define a leading centre (all three grouped as unclear in Table 9.1).

Table 9.1 Publication status and characteristics of included studies

Study characteristics	Approved (column %)	Started at local study site	Of those started:	
			Published (row %)	Not published (row %)
Total	917 (100)	807	419 (52)	388 (48)
Study design				
Randomised controlled trial	408 (45)	355	201 (57)	154 (43)
Non-randomised intervention study	72 (8)	65	33 (51)	32 (49)
Diagnostic study	41 (4)	36	21 (58)	15 (42)
Cohort study	23 (2)	19	8 (42)	11 (58)
Case-control study	6 (1)	6	3 (50)	3 (50)
Cross-sectional study	42 (5)	40	16 (40)	24 (60)
Uncontrolled study	186 (20)	163	75 (46)	88 (54)
Laboratory study	138 (15)	122	61 (50)	61 (50)
Health services research	1 (<1)	1	1 (100)	0
		Pearson χ^2 (df 8) = 10.173, p = 0.253		
Study size				
Size ≥ median of 120	449 (49)	391	224 (57)	167 (43)
Size < median of 120	429 (47)	379	177 (47)	202 (53)
Unclear	39 (4)	37	18 (49)	19 (51)
		Pearson χ^2 (df 2) = 8.808, p = 0.012		
Collaboration				
Single-centre study	383 (42)	340	159 (47)	181 (53)
Multi-centre study	534 (58)	467	260 (56)	207 (44)
		Pearson χ^2 (df 1) = 6.257, p = 0.012		
Only multi-centre studies:				
International	310 (58)	276	173 (63)	103 (37)
Domestic	221 (41)	189	87 (46)	102 (54)
Unclear	3 (<1)	2	0	2 (100)
		Pearson χ^2 (df 2) = 15.124, p = 0.00052		
Leading centre:				
Local	83 (15)	76	41 (54)	35 (46)
Other	448 (84)	388	218 (56)	170 (44)
Unclear[a]	3 (<1)	3	1 (33)	2 (67)
		Pearson χ^2 (df 2) = 0.74, p = 0.691		

(continued)

Table 9.1 (continued)

Study characteristics	Approved (column %)	Started at local study site	Of those started:	
Funding (as stated in protocol)				
Commercial	422 (46)	368	203 (55)	165 (45)
Non-commercial	140 (15)	131	75 (57)	56 (43)
No funding stated	355 (39)	308	141 (46)	167 (54)
		Pearson χ^2 (df 2) = 7.695, p = 0.021		
Only commercially funded studies:				
Sponsor involved	362 (86)	318	182 (57)	136 (43)
Sponsor not involved	60 (14)	50	21 (42)	29 (58)
		Pearson χ^2 (df 1) = 4.053, p = 0.044		

[a]For two studies (one international, one domestic), the leading centre was not determined at the time of REC submission and for one international study there was no intention to define a leading centre

Commercial funding was present in 422 studies (46%) according to protocol information (Table 9.1). In 60 of those, the sponsor provided study drugs or other material, but was not involved in study conduct otherwise. Information on non-commercial funding was given for 140 studies (15%), including applications for funding by the German Research Foundation (Deutsche Forschungsgemeinschaft) in 51 studies and the Federal Government (Ministry of Education and Research/ Bundesministerium für Bildung und Forschung; Federal Ministry of Health/ Bundesministerium für Gesundheit) in 26 studies.

9.3.2 Course of Studies and Publication Outcome

In the survey, we obtained responses for 825 of 917 approved studies (response rate 90%). Including information from other sources, the project status could be determined for 856 studies (93%): 807 (88%) were started at the local study site and 110 (12%) were not started, either locally or in all study centres (Fig. 9.1). Of the 807 initiated studies, 576 (71%) were completed according to protocol, 128 (16%) discontinued and 42 (5%) still ongoing at the time of our study. The latter included studies that were still recruiting participants (n=23), were ongoing after completed recruitment (n=12) or were ongoing after completed data collection (n=7). For 61 (8%) there was no information about current status and we assumed that they had been started, at least.

We identified 782 full publications that corresponded to 419 of the 807 studies. The year of publication ranged from 2000 to 2011. Consequently, the overall publication proportion was 52% (95% CI: 0.48–0.55). The median number of publications per study was 1 and the range was 1–56. Of the 807 initiated studies, 135 (17%) had more than one corresponding publication. Of note, one laboratory study was still ongoing 8 years after ethical approval and had yielded a total of 56 publications until then. In the 770 initiated studies with information about the number of participants (not available for 37), it was planned to recruit at least 298,242 study participants overall (i.e. at all study sites). Of those, 178,254 (60%) participants had their data reported in publications corresponding to the 419 study protocols. In turn, 119,988 (40%) persons participated in the 388 studies that ultimately remained unpublished.

The publication proportion ranged from 40% (95% CI: 0.25–0.57) in cross-sectional studies to 58% in diagnostic studies (95% CI: 0.41–0.74) (Table 9.1). However, the differences by study design did not reach statistical significance ($p=0.253$). In a post-hoc analysis we combined randomized and non-randomized interventional studies; the publication proportion was 56% (95% CI: 0.51–0.61). In contrast, in observational studies (combining cohort, cross-sectional and case-control studies) it was 42% (95% CI: 0.29–0.54). We further analysed whether study size, single or multi-centre status and type of funding were associated with full publication. Larger studies and multi-centre studies were more likely to be published than smaller studies and single-centre studies, respectively (both comparisons: $p=0.012$) (Table 9.1).

In the group of multi-centre studies, the publication proportion of international studies (63%; 95% CI: 0.57–0.68) was higher than of domestic studies (46%; 95% CI: 0.39–0.53; $p=0.00052$). Studies with any funding declared in the protocol (56%; 95% CI: 0.51–0.60) were more often published than studies without (46%; 95% CI: 0.40–0.52; $p=0.021$). Thirty-two (63%) of the 51 studies funded by the German Research Foundation and 12 (46%) of 26 studies funded by the federal government were published.

Of the 419 studies with subsequent publications, evidence of commercial funding was present in the protocols of 203 (48%) and in the corresponding publications of 205 (49%) (Table 9.2). For most of these studies (339; 81%), information on presence or absence of commercial funding was in agreement between protocol and publications. Cohen's kappa was 0.62 (95% CI: 0.54–0.69). However, in 80 (19%) comparisons the funding status did not match: Commercial funding stated in the protocol was not reported in any of the corresponding publications for 39 studies. In turn, commercial funding reported in publications was not stated in the protocol for 41 studies (Table 9.2). Consequently, the ratio of counts of discordant pairs (McNemar odds ratio) was 1.05 (95% CI: 0.66–1.67).

Analogously, evidence of non-commercial funding was present in the protocols of 75 (18%) studies and in corresponding publications of 147 (35%) (Table 9.3). For most of the 419 studies (315; 75%), information on presence or absence of non-commercial funding was in agreement between protocol and publications. Cohen's kappa was 0.39 (95% CI: 0.28–0.49). In 104 (25%), the non-commercial funding status did not match: Noncommercial funding stated in the protocol was not reported

Table 9.2 Funding status in protocols and corresponding publications—commercial funding

		Information in publication, number of studies		
		Yes	No	Total
Information in protocol, number of studies	Yes	164	39	203
	No	41	175	216
	Total	205	214	419

Table 9.3 Funding status in protocols and corresponding publications—non-commercial funding

		Information in publication, number of studies		
		Yes	No	Total
Information in protocol, number of studies	Yes	59	16	75
	No	88	256	344
	Total	147	272	419

For protocols with two or more corresponding publications we regarded funding status as reported if it was found in at least one publication

For publications with both commercial and non-commercial funding, both components were compared separately

in publications for 16 studies, and non-commercial funding reported in publications was not stated in the protocol for 88 studies (Table 9.3); the McNemar odds ratio was 5.50 (95 % CI: 3.21–10.04). In 40 publications (and none of the protocols) there was a statement of both commercial and non-commercial funding.

The predominant language of the publications was English: 367 (88 %) studies were published in English and 25 (6 %) in German. This predominance was found in both international and domestic studies, as well as multi and single centre studies (Table 9.4).

We analysed whether local investigators (i.e. those submitting to the REC) were authors of subsequent publications. In 259 (62 %) of the published 419 studies, local investigators were (co-)authors of at least one corresponding publication (Table 9.4). All but one publication from single-centre studies were authored by a local investigator. In this one publication, an expanded European data set was reported and the local investigator was acknowledged. Publications of 101 (39 %) multicentre studies were authored by a local investigator. In the subgroup of international multi-centre studies this proportion was 34 % (Table 9.4). In multi-centre studies led by the local centre, the local investigators were authors in most studies (35; 85 %), but less often (65; 30 %) if the study was led by another centre.

9.4 Discussion

We analysed clinical research projects approved by a German REC over 3 years, focusing on their publication outcome and the consistency of reporting in aspects such as funding. Only about half of the clinical studies that started recruiting

Table 9.4 Language of publication(s)

Collaboration	Number of published studies (column %)	Median of number of publications (Range)	Number of studies published in English (row %)	Number of studies published in German (row %)	Number of studies published in English and German (row %)	Number of studies with publications (co-) authored by local investigator[a]
Total	419 (100)	1 (1–56)	367 (88)	25 (6)	27 (6)	259 (62)
Single-centre study	159 (38)	1 (1–18)	129 (81)	16 (10)	14 (9)	158 (99)
Multi-centre study	260 (62)	1 (1–56)	238 (92)	9 (3)	13 (5)	101 (39)
International	*173 (67)*	*1 (1–56)*	*163 (94)*	*2 (1)*	*8 (5)*	*58 (34)*
Domestic	*87 (33)*	*1 (1–9)*	*75 (86)*	*7 (8)*	*5 (6)*	*43 (49)*
Leading centre: Local	*41 (16)*	*1 (1–56)*	*30 (73)*	*5 (12)*	*6 (15)*	*35 (85)*
Leading centre: Other	*218 (84)*	*1 (1–9)*	*207 (95)*	*4 (2)*	*7 (3)*	*65 (30)*
Unclear	*1*	*1*	*1 (100)*	*0*	*0*	*1 (100)*

[a] In a study with more than 1 publication local investigators were considered authors if they were authors of at least 1 publication

participants were published as full articles about 8–10 years later. Study design was not associated with full publication. Multicentre status, presence of an international collaboration, large sample size, declared study funding and involvement of sponsor as stated in the protocol were positively associated with subsequent publication.

The Helsinki Declaration of the World Medical Association emphasizes that both authors and publishers of scientific research have ethical obligations and that negative and inconclusive results should be made publicly available, as is the case for positive results (WMA 2013). Our study confirms earlier evidence that the underreporting of clinical research is still prevalent (Dwan et al. 2011). It is sometimes put forward that more rigorous studies (e.g. randomised trials) will be published eventually while studies conducted with less methodological rigour may remain "in the file drawer". In our cohort, study design was not associated with full publication; 43 % of randomised trials had not been published.

It must be of concern that sizeable proportions of studies remain unpublished. Withholding research results pose several ethical problems since participants consent on the premise of contributing to the advancement of medical knowledge and considerable research resources are invested without any benefit in return. In our study, research results of almost 120,000 study participants remained hidden. Not only are patients who are willing to contribute to medical progress betrayed, but also public funds wasted. For instance, 19 of the 51 studies (37 %) funded by the German Research Foundation and 14 of the 26 studies (54 %) funded by the German federal government remained unpublished. Furthermore, non-publication and selective reporting of research results have an impact on the scientific knowledge. For instance, the conclusions of systematic reviews may be biased (Hopewell et al. 2009).

Information on sources of funding is important to appraise the validity of a study's results. It has been shown that commercially funded studies are more likely to produce favourable results and conclusions than those sponsored by other sources (Lundh et al. 2012). Although this information was consistent for most published studies, it is of concern that, firstly, for several studies with commercial funding or non-commercial funding, this information was omitted in the publications and, secondly, that funding sources are not always disclosed to the REC (provided that they are known at the time of submission). The discrepancy regarding funding information is consistent with our earlier finding in a sub-sample of randomised trials from the same cohort: There were important discrepancies in the eligibility criteria for trial participants between protocols and publications (Blümle et al. 2011). The present analysis found that commercial funding information was undisclosed in protocols and publications to the same extent. In contrast, the odds of finding information about non-commercial funding in the publication (but not the protocol) was 5.5 times higher than vice versa. A potential explanation is that industry involvement in the study's planning and conduct had already been determined at the time of writing of most commercially funded protocols, while in non-commercially funded trials, e.g. investigator-initiated trials, funding requests might be pending at this stage and consequently no funding information added to the protocol. Another reason may be that the publishing journals have strict policies for disclosure that incite investigators of non-commercially funded trials to disclose their funding sources more frequently.

Unsurprisingly, our results also show that most studies are published in English, even if the studies are domestic, multi- or single-centre studies with funding from a non-anglophone country, such as Germany. Given that language barriers continue to exist, in particular if new knowledge is to be transferred from research into practice, this must be of great concern. Likely, a sizable part of the healthcare communities not speaking English will not benefit from research findings reported in English language; concomitant efforts to provide translations (e.g. of summaries) are therefore needed (Haße and Fischer 2001; von Elm et al. 2013).

The problem of poorly reported or unreported study results has long been recognised, but is by far not resolved. Clinical trial registries can help to improve transparency and to inform patients, physicians and researchers about planned, ongoing and completed studies (Jena et al. 2010). However, prospective registration is not mandatory for all types of clinical studies and the regulations differ between countries. In the United States "applicable clinical trials", such as those on drugs, biological products and devices, have to be registered since 2007 (http://clinicaltrials.gov/ct2/manage-recs/fdaaa). In the European Union, clinical drug trials submitted to the European Medicine Agency (EMA) are registered in the EudraCT database, but only part of the information is open to the general public.

In Germany, trial registration is still optional and had not yet been introduced at the time of REC approval of the included trials. Therefore, we did not focus on this aspect in the present study. Analyses of more recent research will be able to address the impact of trial registration on publication outcomes more thoroughly.

The lack of access to key data of clinical trials has been put on the agenda of science policy makers and the public again by the recent "All Trials" initiative. This international initiative calls on governments, regulators and research bodies to implement measures to achieve that "all trials past and present should be registered, and the full methods and the results reported" (http://www.alltrials.net). Another recent effort called Restoring Invisible and Abandoned Trials (RIAT), calls on funders and investigators to publish or republish studies that were abandoned and left unpublished. The RIAT proposal provides authors with a set of criteria to assist with precise publication and republication of abandoned studies (Doshi et al. 2013; Loder et al. 2013). Our empirical data underpins these efforts suggesting that the magnitude of underreporting has not diminished yet, despite joint large-scale initiatives such as trial registration.

Our comprehensive literature search employed several databases and was complemented by an investigator survey with a high response rate. We are confident that most full articles corresponding to the included study protocols could be identified. Despite these efforts, we cannot rule out that some were missed. Consequently, the publication proportion may be underestimated. On the other hand, we regarded several discontinued studies with published preliminary results as fully published, which could be perceived as an overestimation of the publication proportion. We excluded conference abstracts and other so-called "grey literature" because those publications are often not indexed in electronic databases (in particular, abstracts of smaller conferences). Many of them are not found even by extensive literature searches and resulting estimates of publication outcome would therefore likely be incomplete or even biased. Further, we had to rely on several arbitrary definitions when extracting data and classifying studies. Since we included all types of studies submitted for ethical approval, we classified protocols by study design using a classification scheme that had proven useful in previous studies (von Elm et al. 2008; Blümle et al. 2008). Arguably, other criteria could have been used. For clinical trials, we decided against using the phase I to IV classification since it was not applied consistently in the included protocols. Alternative definitions would have been possible also for other study variables. However, given that all variables were defined a priori we are confident that our choices did not lead to any systematic error in our analyses. Clearly, it would be interesting to analyse more recent study protocols, as the quality of protocols and publications and the practices of scientific reporting change over time. In particular, trial registration has been introduced more widely since then. However, sufficient time must have elapsed before the ultimate fate of studies with regard to completion and publication can be determined. The obvious dilemma is that including more recent protocols would have left insufficient time for studies to be completed and results to be published (Scherer et al. 2007). In our sample, about 5 % of studies were still ongoing 8–10 years after ethical approval. We chose to analyse the period from REC approval to publication because reliable data for both these time points were available. An estimate of the time elapsed between completion of the study (e.g. end of data collection) and publication would have been more meaningful. However, such information was not included regularly in study reports or REC files.

It would also have been interesting to investigate the reasons for non-publication. However, based on our prior experience with approaching local investigators for empirical research, we deemed that it is not feasible in a postal survey (in particular up to 10 years later) as it implies asking sensitive questions and likely would have influenced the response rate negatively. In fact, in many cases, non-publication of research has to do with poor project management, disagreements in research groups or other unforeseen events, and it is unlikely that trialists would have disclosed such circumstances in a survey.

We used a sample of studies conducted in various disciplines at a large German university. Many were multi-centric, international or both and studies could be included without seeking the trialists' consent. We are therefore confident that our results have some external validity in similar clinical research environments in other high-income countries.

9.5 Conclusions

In a large unselected sample of clinical research projects approved by a German research ethics committee, only about half of the started studies were published. In addition, 16 % of the started studies were discontinued. Crucial information such as study funding differed between protocols and publications in about 20 % of published trials. If only part of the accumulated research data are accessible for those potentially interested, scarce research resources are wasted. Furthermore, health care professionals and patients cannot make decisions based on all the available evidence and other researchers may build future projects on an incomplete or even biased database.

Acknowledgements We are grateful to the research ethics committee of the Albert-Ludwigs-Universität Freiburg/Germany, its administrative staff, and the local investigators who responded to our survey. We also thank Alexander Hellmer and Florian Volz for help with extraction of protocol data, and Patrick Oeller for updating the literature search.

References

Altman, D.G. 1991. *Practical statistics for medical research*. Boca Raton/London/New York/Washington, DC: Chapman and Hall/CRC.

Blümle, A., G. Antes, M. Schumacher, H. Just, and E. von Elm. 2008. Clinical research projects at a German medical faculty: Follow-up from ethical approval to publication and citation by others. *Journal of Medical Ethics* 34: e20.

Blümle, A., J.J. Meerpohl, G. Rucker, G. Antes, M. Schumacher, and E. von Elm. 2011. Reporting of eligibility criteria of randomised trials: Cohort study comparing trial protocols with subsequent articles. *BMJ* 342: d1828.

Chalmers, I., and P. Glasziou. 2009. Avoidable waste in the production and reporting of research evidence. *Lancet* 374: 86–89.

Decullier, E., V. Lheritier, and F. Chapuis. 2005. Fate of biomedical research protocols and publication bias in France: Retrospective cohort study. *BMJ* 331: 19.

Dickersin, K., and I. Chalmers. 2010. Recognising, investigating and dealing with incomplete and biased reporting of clinical research: From Francis Bacon to the World Health Organisation. *JLL Bulletin: Commentaries on the history of treatment evaluation.* http://www.jameslindlibrary.org. Accessed 13 Feb 2015.

Dickersin, K., Y.I. Min, and C.L. Meinert. 1992. Factors influencing publication of research results. Follow-up of applications submitted to two institutional review boards. *JAMA* 267: 374–378.

Doshi, P., K. Dickersin, D. Healy, S.S. Vedula, and T. Jefferson. 2013. Restoring invisible and abandoned trials: A call for people to publish the findings. *BMJ* 346: f2865.

Dwan, K., D.G. Altman, L. Cresswell, M. Blundell, C.L. Gamble, and P.R. Williamson. 2011. Comparison of protocols and registry entries to published reports for randomised controlled trials. *Cochrane Database of Systematic Reviews* 1: MR000031.

Easterbrook, P.J., J.A. Berlin, R. Gopalan, and D.R. Matthews. 1991. Publication bias in clinical research. *Lancet* 337: 867–872.

Haße, W., and R. Fischer. 2001. Englisch in der Medizin: Der Aus- und Weiterbildung hinderlich. *Dtsch Arztebl* 98: 3100–3102.

Higgins, J.P.T., and S. Green, editors. 2011. *Cochrane handbook for systematic reviews of interventions version 5.1.0.* The Cochrane Collaboration. http://www.cochrane-handbook.org. Accessed 13 Feb 2015.

Hopewell, S., K. Loudon, M.J. Clarke, A.D. Oxman, and K. Dickersin. 2009. Publication bias in clinical trials due to statistical significance or direction of trial results. *Cochrane Database of Systematic Reviews* 1: MR000006.

Jena, S., G. Dreier, G. Antes, and M. Schumacher. 2010. Clinical trials. More transparency in EU testing? *European Biotechnology News* 9(5–6): 28–30.

Loder, E., F. Godlee, V. Barbour, and M. Winker. 2013. Restoring the integrity of the clinical trial evidence base. *BMJ* 346: f3601.

Lundh, A., S. Sismondo, J. Lexchin, O.A. Busuioc, and L. Bero. 2012. Industry sponsorship and research outcome. *Cochrane Database of Systematic Reviews* 12: MR000033.

Rosenthal, R. 1979. The "file drawer problem" and tolerance for null results. *Psychological Bulletin* 86: 638–641.

Rothman, K.J., S. Greenland, and T.L. Lash. 2008. *Modern epidemiology*, 3rd ed. Philadelphia: Wolters Kluwer Health, Lippincott, Williams & Wilkins.

Scherer, R.W., P. Langenberg, and E. von Elm. 2007. Full publication of results initially presented in abstracts. *Cochrane Database of Systematic Reviews* 2: MR000005.

Song, F., S. Parekh, L. Hooper, Y.K. Loke, J. Ryder, and A.J. Sutton, et al. 2010. Dissemination and publication of research findings: An updated review of related biases. *Health Technology Assessment* 14: iii, ix–xi, 1–193.

Stern, J.M., and R.J. Simes. 1997. Publication bias: Evidence of delayed publication in a cohort study of clinical research projects. *BMJ* 315: 640–645.

Strech, D. 2012. Normative arguments and new solutions for the unbiased registration and publication of clinical trials. *Journal of Clinical Epidemiology* 65: 276–281.

von Elm, E., A. Rollin, A. Blümle, K. Huwiler, M. Witschi, and M. Egger. 2008. Publication and non-publication of clinical trials: Longitudinal study of applications submitted to a research ethics committee. *Swiss Medical Weekly* 138: 197–203.

von Elm, E., P. Ravaud, H. MacLehose, L. Mbuagbaw, P. Garner, J. Ried, et al. 2013. Translating Cochrane reviews to ensure that healthcare decision-making is informed by high-quality research evidence. *PLoS Medicine* 10: e1001516.

World Medical Association (WMA). 2013. *Declaration of Helsinki: Ethical principles for medical research involving human subjects.* 64th WMA General Assembly. Fortaleza, Brazil. http://www.wma.net/en/30publications/10policies/b3. Accessed 15 Feb 2015.

Chapter 10
Do Editorial Policies Support Ethical Research? A Thematic Text Analysis of Author Instructions in Psychiatry Journals

Daniel Strech, Courtney Metz, and Hannes Kahrass

Abstract According to the Declaration of Helsinki and other guidelines, clinical studies should be approved by a research ethics committee and seek valid informed consent from the participants. Editors of medical journals are encouraged by the ICMJE and COPE to include requirements for these principles in the journal's instructions for authors. This study assessed the editorial policies of psychiatry journals regarding ethics review and informed consent. The information given on ethics review and informed consent and the mentioning of the ICMJE and COPE recommendations were assessed within author's instructions and online submission procedures of all 123 eligible psychiatry journals. While 54 and 58 % of editorial policies required ethics review and informed consent, only 14 and 19 % demanded the reporting of these issues in the manuscript. The TOP-10 psychiatry journals (ranked by impact factor) performed similarly in this regard. Only every second psychiatry journal adheres to the ICMJE's recommendation to inform authors about requirements for informed consent and ethics review. Furthermore, we argue that even the ICMJE's recommendations in this regard are insufficient, at least for ethically challenging clinical trials. At the same time, ideal scientific design sometimes even needs to be compromised for ethical reasons. We suggest that features of clinical studies that make them morally controversial, but not necessarily unethical, are analogous to methodological limitations and should thus be reported explicitly. Editorial policies as well as reporting guidelines such as CONSORT should be extended to support a meaningful reporting of ethical research.

Originally published as: Strech D, Metz C, Knüppel H. Do Editorial Policies Support Ethical Research? A Thematic Text Analysis of Author Instructions in Psychiatry Journals. PLoS ONE. 2014;9(6):e97492. doi:10.1371/journal.pone.0097492.

D. Strech (✉) • C. Metz • H. Kahrass
Institute for History, Ethics and Philosophy of Medicine, Hannover Medical School,
Carl-Neuberg-Str. 1, Hannover 30625, Germany
e-mail: strech.daniel@mh-hannover.de

10.1 Introduction

According to the Declaration of Helsinki, research studies should (1) be approved by an independent research ethics committee (REC) and (2) seek informed consent (IC) from the participants (WMA 2013). These principles have in turn been addressed by the International Committee of Medical Journal Editors (ICMJE) and the Committee on Publication Ethics (COPE). Both groups publish core requirements for editing and reporting research findings. For example the ICMJE state in their recommendations (previously known as uniform requirements for manuscripts) that "the requirement for informed consent should be included in the journal's Instructions for Authors. When informed consent has been obtained, it should be indicated in the published article." The COPE code of conduct asks editors to ensure that reports of clinical trials cite compliance with the Declaration of Helsinki (DoH), Good Clinical Practice, and other relevant guidelines on safeguarding participants. Editors are encouraged by the ICMJE and COPE to apply and distribute these guidelines (COPE 2011; ICMJE 2013). Consequential responsibilities of journal editors have been widely discussed (Amdur and Biddle 1997; Angelski et al. 2012; Brackbill and Hellegers 1980; Fernandez and Garcia 2005; McDonald 1985; Wagner and Kleinert 2010).

However, empirical data from several studies throughout the last two decades suggest insufficient reporting of ethics review approvals and IC procedures in peer-reviewed articles and metaanalysis (Asai and Shingu 1999; Bauchner and Sharfstein 2001; Finlay and Fernandez 2008; Matot et al. 1998; Olde Rikkert et al. 1996; Olson and Jobe 1996; Stocking et al. 2004; Weil et al. 2002; Yank and Rennie 2002). Weil and colleagues demonstrated that only 52 % of the articles in paediatric journals reported ethical approval and one in seven studies had not undergone REC review (Weil et al 2002).

A few studies have assessed journals' instructions to authors on the reporting of ethical issues, but none has done so in the field of psychiatry, and no study so far has investigated both the instructions given to authors on the journals' websites and those given during the submission process (Amdur and Biddle 1997; Bavdekar et al. 2008; Navaneetha 2011; Rowan-Legg et al. 2009). Furthermore, editorial policies on more specific reporting of ethical approval or informed consent have not yet been assessed systematically. For example, more specific reporting might be expected with regard to how the capacity to give informed consent was assessed in patients with Alzheimer or schizophrenia (Miller et al. 1999; Stocking et al. 2004). More specific reporting on ethics review might be expected with respect to the justification of studies in which patients, for example, (i) receive placebos, (ii) are withdrawn from standard medication, (iii) undergo "wash out" phases or (iv) are administered a challenge agent (Miller et al. 1999; Riedel et al. 2012).

The objective of this study was to assess the editorial policies of a representative sample of psychiatry journals on the reporting of ethics review and informed consent in original research papers. Furthermore, this study assessed whether and how psychiatry journals refer to international guidelines on publication ethics.

10.2 Methods

Based on Journal Citation Reports (Thomson Reuters 2013) from 2011 we identified 130 journals indexed in the subject category "psychiatry". We further specified a subsample of 10 psychiatry journals with the highest impact factor ("TOP-10"). We restricted our analysis to journals published in English or German. We accessed the "author's instructions" or similar texts on the journals' websites between July and August 2012. We further accessed the instructions given during the online submission procedure in January 2013. The online submission procedures were entered by a fake submission of an "original paper", or a "clinical research" or "clinical trial" paper. All PDFs or website texts were downloaded using WinHTTrack 3.46-1 for documentation. The membership of all journals of COPE or ICMJE was checked on the respective web pages (www.publicationethics.org and www.icmje.org) in August 2013.

We assessed if and how the DoH, ICMJE and COPE were mentioned in the author instructions or during the submission procedure. Further we assessed the information given on ethics review, and informed consent. We had three rating options: (1) "not mentioned", (2) "information recommended" or (3) "information required". The rating "information recommended" was applied to moderate wording in the author instructions such as "should" or "we recommend that…" The rating "information required" was applied to strong wording like "authors must…", "we expect authors to…" or "we require authors to…" Particular specifications and requirements on ethics approval and informed consent were extracted and recorded.

Multiple designations of the responsible ethical authority was treated as referring to the same body (ethical review board, ethical review committee, research ethics board or institutional review board). In this text we use the term "research ethics committee" (REC) consistently.

Two authors (HK, CM) independently assessed the editorial policies and then merged their findings. Inconsistent findings were discussed in consultation with a third member of the group (DS).

We calculated frequency data using standard descriptive statistics.

10.3 Results

After excluding 7 journals which were not in English or German, or had no web page, we included 123 journals in our analysis (116 in English and 7 in German).

10.3.1 Information and Requirements Regarding International Guidelines on Publication Ethics

Of the 123 psychiatry journals, 46 (37%) referred to the Declaration of Helsinki in the author instructions or during their online submission process. Sixty-eight (55%) of all journals referred to the ICMJE but of these only 11 (17%) were listed as "following URM" (now: "following ICMJE recommendations") on icmje.org. Conversely, while 28 (33%) of all journals referred to COPE in the author instructions or during their online submission process, 62 (50%) were indicated as signed up to COPE on publicationethics.org.

From the TOP-10 psychiatry journals (ranked by impact factor) 20% referred to the Declaration of Helsinki, 90% to the ICMJE's URMs (now: ICMJE recommendations), and 20% to COPE. Fifty percent were listed as "following URM" (now: "following ICMJE recommendations") on the ICMJE website and 60% as signed up on the COPE website.

10.3.2 Information and Requirements Regarding Ethics Review

Of the 123 psychiatry journals, 66 (54%) recommended or required ethics review explicitly in their author instructions or during their online submission process, but only 17 (14%) required that REC approval must be mentioned in the manuscript (Table 10.1). Further specifications or additional requirements on the reporting of ethics review information were made by 20 (16%) of all 123 journals. Twelve (10%) required the reporting of the REC's name and seven (6%) an original document for REC approval. No editorial policy asked for justifications from the principal investigator or the ethics review board with respect to particular risks and ethical concerns in the research design (e.g. the need to withdraw a patient's standard medications or administration of a challenge agent that can provoke psychiatric symptoms). The findings for the TOP-10 journals have the same tendency (Table 10.1).

10.3.3 Information and Requirements on Informed Consent

Giving the editor details of the informed consent (IC) procedure was recommended or required by 71 (58%) of all 123 journals, but only 23 (19%) required this information in the manuscript (Table 10.2).

Further specifications or additional requirements on the reporting of informed consent were made by 18 (15%) of the journals: seven (6%) asked for information

Table 10.1 Detailed information about the statements on ethics review in the authors' instructions or during the submission process

Ethics review	Wording examples according to the ratings	All journals N=123	Top-10 journals N=10
Recommended or required ethics review		66 (54%)	5 (50%)
Specified their requirements concerning the statement		27 (22%)	3 (30%)
Publication of the information in the manuscript	"For human or animal experimental investigations, appropriate institutional review board approval is required and should be described in the Methods section of the paper."	17 (14%)	2 (20%)
Required to name the REC	"Manuscripts that involve investigations on human participants must give the name of the ethics committee that approved the study."	12 (10%)	2 (20%)
Required original documents or evidence	"An author must make available all requisite formal and documented ethical approval from an appropriate research ethics committee using humans or human tissue."	7 (6%)	0 (0%)
Required explanation, if there was no REC approval	"State whether institutional review board approval was obtained for the investigation; if it was not, provide an explanation."	3 (2%)	0 (0%)
Required to report exemption or requirements from the REC	"If a study has been granted an exemption from requiring ethics approval, this should also be detailed in the Methods section."	2 (2%)	0 (0%)
Not recommended or required ethics review		57 (46%)	5 (50%)

on how the decision-making capacity of participants was assessed, and 5 (4%) requested to know whether the child's assent was obtained in addition to the informed consent of the child's proxies. Ten journals (8%) asked for the original IC templates to be provided to the editor. The ratings for the TOP-10 journals were similar (see Table 10.2).

Furthermore, the 71 journals that required statements on IC differed in whether and how they demanded particular issues to be addressed in the IC forms. While 58 journals asked authors in a rather general manner to "include a statement in the manuscript that informed consent was obtained" 13 (11%) journals further specified what the IC should include. The most frequent specification was that IC procedures must include a "full explanations of the procedures" (n=9). Others state e.g. that the study subject should be informed about "possible side effects" (n=3), "purpose of the research" (n=1), "the right to decline to participate and to withdraw from the research once participation has begun" (n=1), "prospective research benefits" (n=1), "limits of confidentiality" (n=1), or "incentives for participation" (n=1).

Table 10.2 Detailed information about the statements on informed consent (IC) in the authors instructions or during the submission process

Informed consent	Wording examples according to the ratings	All journals N = 123	Top-10 journals N = 10
Recommended or required IC		71 (58 %)	8 (80 %)
Specified their requirements concerning the statement		34 (28 %)	4 (40 %)
Publication of the information in the manuscript	"Within the Methods section, authors should indicate that 'informed consent' has been appropriately obtained and state the name of the REC, IRB or other body that provided ethical approval."	23 (19 %)	2 (20 %)
Required information on the capacity assessment	"Authors of reports on human studies should include detailed information on the informed consent process, including the method(s) used to assess the subject's capacity to give informed consent, and safeguards included in the study design for protection of human subjects."	7 (6 %)	2 (20 %)
Required information on child's assent	"In the case of children, authors are asked to include information about whether the child's assent was obtained in addition to that of the legal guardian."	5 (4 %)	1 (10 %)
Required original documents or evidence	"An author must make available all requisite formal and documented ethical approval from an appropriate research ethics committee using humans or human tissue, including evidence of anonymisation and informed consent from the client(s) or patient(s) studied."	10 (8 %)	1 (10 %)
Not recommended or required IC		52 (42 %)	2 (20 %)

10.4 Discussion

Several international policies and guidelines aim to improve the adherence to ethical standards and responsible conduct in clinical research and its reporting. For example, the ICMJE and COPE advise medical journal editors to require information about informed consent (IC) procedures and the approval of the local research ethics committee/institutional review board (REC/IRB). A minor but nevertheless striking finding of this study is the inconsistency between the number of journals mentioning one of these organisations (ICMJE = 55 % and COPE = 33 %) in their editorial polices and the number of journals officially registered with these organisations (ICMJE = 9 % and COPE = 50 %). This inconsistency questions the seriousness of mentioning or signing up to these organisations at least for the journals that either mention or have signed up but not both.

The ICMJE recommend that the requirement for informed consent should be included in the journal's instructions to authors, and that the published article should indicate when informed consent has been obtained. While every second editorial policy of the 123 reviewed psychiatry journals recommended or required REC approval (54 %) or IC procedure (58 %) in the author instructions, only a minority of journals explicitly demanded the reporting of these issues in the manuscript (14 % and 19 %). The TOP-10 psychiatry journals (ranked by impact factor) performed similarly in this regard. Against this background it is unsurprising that only a tiny minority of editorial policies asked for the reporting of more detailed information of particulars in the ethics review (e.g. justification of ethically challenging study designs) or in the informed consent procedures (e.g. how informed consent was obtained in participants with impaired decision making or how decision-making capacity was assessed prior to informed consent). Stocking et al. found in a review of trials on Alzheimer disease that only 8 % reported that decision-making capacity was assessed specifically for the reported study and that this assessment was completed before recruitment (Stocking et al. 2004).

We justify in the following paragraphs why editorial policies of psychiatry journals (as well as other general and specialty journals) should require more transparent, more consistent, and more detailed reporting regarding ethical issues of published studies. Insufficient reporting of ethical issues within biomedical research can negatively affect how trustworthy the public judge the biomedical research community to be (Hardin 2006). Public trust in the research community requires evidence that this specific community has qualities such as competence and good will which merit that trust (Tullberg 2007).

Insufficient reporting of ethical issues may not only give the impression to the public but also to the research community itself that the ethical quality of research is judged far less important than its scientific validity. However, designing a study demands both critical reflection on relevant methodological aspects (e.g. randomisation and blinding to minimise the influence of confounding biases) and on ethical issues (e.g. fair selection of, minimising risks for and obtaining valid informed consent from trial participants) (Emanuel et al. 2000). Furthermore, ideal scientific design sometimes needs to be compromised for ethical reasons.

The better established requirement to report standard methodological aspects (e.g. eligibility criteria, blinding, randomization procedures (Schulz et al. 2010)) has two main consequences: First, as a direct consequence it helps editors, reviewers and readers to assess the reliability and validity of the research. Second, as an indirect consequence it signals to future authors the importance of critical reflection on methodological quality in the design and conduct of a study. Likewise, editorial policies should require reporting of pertinent ethical considerations for the following reasons: (A) to allow editors, reviewers and readers to assess the ethical quality of the research, (B) to foster the design and conduct of future studies that meet appropriate standards of ethical research (Miller and Rosenstein 2002), (C) to raise the visibility of ethical research and thereby maintain public trust and (D) to facilitate a discussion and scientific evaluation of current standards and variations in real-life research ethics.

General statements such as "the study was approved by the local IRB" or "informed consent was obtained from all study participants" clearly do not meet the above described rationale for and aims of reporting ethical issues—at least not in research involving patients with disorders that may impair decision-making capacity, such as Alzheimer disease and schizophrenia, nor in research involving interventions that pose ethical concerns (see examples above and in Miller et al. (1999) and Riedel et al. (2012)).

Against the background of the presented empirical findings and normative analysis, and in accordance with former suggestions from Miller et al. (1999) we suggest that features of clinical studies that make them morally controversial, but not necessarily unethical, are analogous to methodological limitations. Editorial policies should be revised to support a meaningful reporting of ethical research. To reach this aim, the current COPE and URM recommendations concerning the reporting of IC and REC approval should also be revised.

Studies that have morally controversial features, such as placebo controls, symptom provocation or deception, might be dismissed as unethical unless the rationale for including such features and details of safeguards to protect research participants from harm or exploitation are explained (Miller and Rosenstein 2002).

Following this line of argumentation and adding the premise that using results of (presumably) unethical studies is (at least) morally doubtful we also recommend in accordance with Weingarten et al. (2004) to include an ethical assessment in systematic reviews of clinical trials. This recommendation should be considered in revisions of manuals for systematic reviews (Cochrane handbook (Higgins and Green 2011) as well as in revisions or extension of reporting guidelines such as CONSORT (Schulz et al. 2010) or PRISMA (Moher et al. 2009).

10.5 Conclusion

Only every second psychiatry journal adheres to the ICMJE's recommendation to inform authors about requirements for informed consent and ethics review. Furthermore, only 14 % and 19 % of all psychiatry journals demanded the reporting of these issues in the manuscript. The TOP-10 psychiatry journals (ranked by impact factor) performed similarly in this regard. Editors have the opportunity, the right and the competence to support ethical research by (simply) updating their policies on how to report on ethical issues in clinical research.

References

Amdur, R.J., and C. Biddle. 1997. Institutional review board approval and publication of human research results. *JAMA* 277: 909–914.

Angelski, C., C.V. Fernandez, C. Weijer, and J. Gao. 2012. The publication of ethically uncertain research: Attitudes and practices of journal editors. *BMC Medical Ethics* 13: 4.

Asai, T., and K. Shingu. 1999. Ethical considerations in anaesthesia journals. *Anaesthesia* 54: 192–197.

Bauchner, H., and J. Sharfstein. 2001. Failure to report ethical approval in child health research: Review of published papers. *BMJ* 323: 318–319.

Bavdekar, S.B., N.J. Gogtay, and S. Wagh. 2008. Reporting ethical processes in two Indian journals. *Indian Journal of Medical Sciences* 62: 134–140.

Brackbill, Y., and A.E. Hellegers. 1980. Ethics and editors. *Hastings Center Report* 10: 20–24.

Committee on Publication Ethics (COPE). 2011. *International standards for editors and authors.* http://publicationethics.org/international-standards-editors-and-authors. Accessed 13 Feb 2015.

Emanuel, E.J., D. Wendler, and C. Grady. 2000. What makes clinical research ethical? *JAMA* 283: 2701–2711.

Fernandez, E., and A.M. Garcia. 2005. The value of the letters to the editor. *Gaceta Sanitaria* 19: 354–355.

Finlay, K.A., and C.V. Fernandez. 2008. Failure to report and provide commentary on research ethics board approval and informed consent in medical journals. *Journal of Medical Ethics* 34: 761–764.

Hardin, R. 2006. *Trust*. Cambridge: Polity.

Higgins, J.P.T., and S. Green, editors. 2011. *Cochrane handbook for systematic reviews of interventions version 5.1.0*. The Cochrane Collaboration. http://www.cochrane-handbook.org. Accessed 13 Feb 2015.

International Committee of Medical Journal Editors (ICMJE). 2013. *Recommendations for the conduct, reporting, editing, and publication of scholarly work in medical journals*. http://www.icmje.org/recommendations. Accessed 13 Feb 2015.

Matot, I., R. Pizov, and C.L. Sprung. 1998. Evaluation of institutional review board review and informed consent in publications of human research in critical care medicine. *Critical Care Medicine* 26: 1596–1602.

McDonald, A. 1985. Ethics and editors: When should unethical research be published? *CMAJ* 133: 803–805.

Miller, F.G., and D.L. Rosenstein. 2002. Reporting of ethical issues in publications of medical research. *Lancet* 360: 1326–1328.

Miller, F.G., D. Pickar, and D.L. Rosenstein. 1999. Addressing ethical issues in the psychiatric research literature. *Archives of General Psychiatry* 56: 763–764.

Moher, D., A. Liberati, J. Tetzlaff, D.G. Altman, and PRISMA Group. 2009. Preferred reporting items for systematic reviews and meta-analyses: The PRISMA statement. *Annals of Internal Medicine* 151: 264–269. W64.

Navaneetha, C. 2011. Editorial policy in reporting ethical processes: A survey of 'instructions for authors' in international indexed dental journals. *Contemporary Clinical Dentistry* 2: 84–87.

Olde Rikkert, M.G., H.A. ten Have, and W.H. Hoefnagels. 1996. Informed consent in biomedical studies on aging: Survey of four journals. *BMJ* 313: 1117.

Olson, C.M., and K.A. Jobe. 1996. Reporting approval by research ethics committees and subjects' consent in human resuscitation research. *Resuscitation* 31: 255–263.

Riedel, M., S. Leucht, E. Ruther, M. Schmauss, and H.J. Moller. 2012. Critical trialrelated criteria in acute schizophrenia studies. *European Archives of Psychiatry and Clinical Neuroscience* 262: 151–155.

Rowan-Legg, A., C. Weijer, J. Gao, and C. Fernandez. 2009. A comparison of journal instructions regarding institutional review board approval and conflict-of-interest disclosure between 1995 and 2005. *Journal of Medical Ethics* 35: 74–78.

Schulz, K.F., D.G. Altman, and D. Moher. 2010. CONSORT 2010 statement: Updated guidelines for reporting parallel group randomised trials. *PLoS Medicine* 7: e1000251.

Stocking, C.B., G.W. Hougham, A.R. Baron, and G.A. Sachs. 2004. Ethics reporting in publications about research with Alzheimer's disease patients. *Journal of the American Geriatrics Society* 52: 305–310.

Thomson, Reuters. 2013. *Journal citation reports*. http://thomsonreuters.com/journal-citation-reports. Accessed 13 Feb 2015.

Tullberg, J. 2007. Trust – The importance of trustfulness versus trustworthiness. *The Journal of Socio-Economics* 37: 2059–2071.

Wager, E., and S. Kleinert. 2010. Responsible research publication: International standards for authors. A position statement developed at the 2nd world conference on research integrity. In *Promoting research integrity in global enviromental*, ed. T. Mayer and N. Steneck, 309–316. Singapore: World Scientific Publishing.

Weil, E., R.M. Nelson, and L.F. Ross. 2002. Are research ethics standards satisfied in pediatric journal publications? *Pediatrics* 110: 364–370.

Weingarten, M.A., M. Paul, and L. Leibovici. 2004. Assessing ethics of trials in systematic reviews. *BMJ* 328: 1013–1014.

World Medical Association (WMA). 2013. *Declaration of Helsinki: Ethical principles for medical research involving human subjects*. 64th WMA General Assembly. Fortaleza, Brazil. http://www.wma.net/en/30publications/10policies/b3. Accessed 15 Feb 2015.

Yank, V., and D. Rennie. 2002. Reporting of informed consent and ethics committee approval in clinical trials. *JAMA* 287: 2835–2838.

Part III
Improving Common Domains of Research Governance

Chapter 11
Ensemble Space and the Ethics of Clinical Development

Jonathan Kimmelman and Spencer Phillips Hey

Abstract For a drug to be clinically useful, physicians must know the optimal conditions for its application, including dose, timing of administration, and route of delivery. Typically, these conditions are discovered over the course of early phase clinical testing. In what follows, we describe a conceptual tool—"ensemble space"—for understanding scientific decision-making in the early phases of clinical development. Briefly, we liken each condition to a dimension that can assume an optimal value for clinical utility. Early stages of intervention development can thus be described as a process of exploring a multi-dimensional landscape of conditions, with the aim of identifying necessary and sufficient conditions (i.e. effective "intervention ensembles") to unlock the clinical utility of a drug. We then show how the concept of ensemble space can be used to address and resolve perennial scientific and ethical debates in clinical development, such as how aggressively to design early phase studies, when to initiate randomized trials, and reporting requirements for early phase studies. We close by discussing some limitations of the concept of ensemble space.

11.1 Introduction

Misused medical interventions are harmful. Adverse drug reactions are between the 4th and 6th leading cause of death, and over 40,000 children are hospitalized each year due to preventable adverse events (Woods et al. 2005). It is only through the arduous work of clinical development that new medical interventions—which include drugs, devices, vaccines, and procedures—are transformed from poisons to therapies.

Clinical development is, however, a burdensome, expensive, and failure-prone undertaking, rife with ethical controversy. At what point is it ethical to expose patients to an unproven substance? How rapidly should early phase trials dose escalate?

J. Kimmelman (✉) • S.P. Hey
Biomedical Ethics Unit, McGill University, 3647 Peel St., Montreal, QC H3A 1X1, Canada
e-mail: jonathan.kimmelman@mcgill.ca

At what point is it acceptable to substitute an unproven substance for standard of care in a randomized trial? What kinds of patients should be eligible for early phase trial participation? What level of non-therapeutic research risk is acceptable in trials of novel drugs? At what point is it acceptable to expose children or incapacitated patients to an unproven drug?

Answers to these and many other questions require an accurate conceptual understanding of clinical translation—what precisely it is, and how can it accomplish these objectives. This chapter is a contribution to the larger project of using empirical methods and philosophical analysis to better understand the epistemic and ethical dimensions of clinical translation.[1]

In what follows, we present a conceptual tool—ensemble space—that can help in negotiating key scientific and ethical challenges in the early phases of research.[2] After describing some applications, we explore some unresolved problems with the concept of ensemble space and point to avenues through which these problems might be resolved.

11.2 The Risk/Benefit Landscape of Early Phase Research

The process of clinical development can be broadly divided into two phases: the early phase (which typically refers to phase 1 and most phase 2 studies), and the late phase (which refers to some phase 2 trials and almost all phase 3 and 4 trials). Although this division originates in regulatory policy, it neatly captures two epistemically-distinct research activities:

1. In the early phase, researchers set out to explore conditions for unlocking the clinical utility of an intervention. Indeed, the therapeutic activity of any medical intervention is conditional on its being combined and coordinated with a set of materials, practices, and knowledge. For example, drugs are only useful when delivered at the right dose, to the appropriate populations, on a designated schedule, with proper safety measures, and aimed at the appropriate clinical outcomes. We call these sets of coordinated materials and practices "intervention ensembles" (Kimmelman 2012).
2. In the late phase, researchers must decisively confirm that an intervention ensemble has therapeutic utility. New intervention ensembles that are validated in confirmatory trials are then taken into clinical practice.

The key task of the early phase is to identify the optimal components of an intervention ensemble—i.e., those combinations of materials and practices that produce the strongest signal of clinical activity. Many of these components scale with therapeutic utility. For example, low doses of a drug may be insufficient to affect disease

---

[1] Other contributions and activities related to this project can be found at our research group's web page, www.translationalethics.com, and also in many of the references contained here.
[2] We have presented this concept more formally elsewhere. See Hey and Kimmelman 2014.

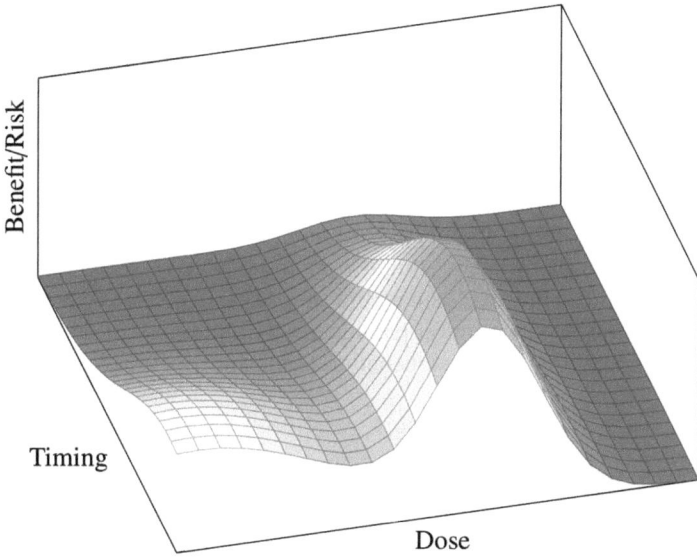

Fig. 11.1 Benefit/risk landscape of ensemble space. The x- and y-axes (dose and timing, respectively) scale with increased risk or reduction in benefit. The z-axis represents benefit signal: *green* is positive signal, *red* is negative signal, and *white* is a neutral outcome (little harm, little benefit signal)

processes, moderate doses may strike a balance between side effects and clinical activity, and high doses may cause intolerable side-effects. Similarly, a drug may not produce any benefit if it is administered too long after the disease event (e.g. stroke) has occurred, produce only modest benefit if used immediately after the event, and yet offer significant benefit if used protectively (i.e., in advance of the disease event).

Given that they scale with risk and benefit, these two ensemble components—dose-response and timing-response—can be represented as dimensions in a 3-dimensional "ensemble space,"[3] illustrated in Fig. 11.1. The two ensemble dimensions are the x and y-axes; the vertical z-axis is a measure of benefit/risk (e.g., signal of clinical activity/adverse event profile).[4]

At the inception of a drug's development, researchers might be aware of the existence of some key dimensions, such as dose, schedule, or diagnostic criteria. They often have some knowledge about the general relationship between dimensions and risk. For example, absent compensatory therapeutic activity, toxicity generally increases with dose; delivering a cell therapy closer to a sensitive anatomical

---

[3] The concept has some relationship to those used to describe identification of optimal dosing in early phase combination therapy studies. See Piantadosi 2013.

[4] Although we normally refer to "risk/benefit," this ratio is inverted in our representation in order to accord with the visual intuition that "higher" values on the z-axis represent more positive outcomes.

region is more dangerous than delivering farther from the sensitive region. Researchers may also have reliable knowledge about risk/benefit at certain positions on the landscape. For instance, they might know that, below a certain dose, the drug is almost certainly inactive (but minimally toxic), and beyond a certain dose the drug is almost certainly toxic. Yet researchers may have little awareness or understanding of other key dimensions at the outset of clinical development—such as delivery or co-interventions—that will need to be explored. And, by definition, they know little to nothing about the "location" of the peaks/optima. Indeed, this information is precisely what is sought over the course of early phase research. Ensemble space thus helps to elucidate this aim of early phase research: it is a systematic exploration of intervention parameters with the goal of achieving a more complete understanding of the risk/benefit relationship.

11.3 Critical Points on the Landscape

The ensemble space representation also helps to make explicit that researchers need to resolve three "critical points" along each dimension: the minimal effective value (e.g., the lower boundary of active dose and effective timing), the approximately peak value (e.g., optimal dose and timing), and the maximum tolerated value (e.g., the upper boundary of acceptable toxicity and time of futility).

There are clinical, scientific, and policy reasons that these critical points should be determined during exploratory stages. The clinical reasons stem from the fact that physicians often adjust intervention ensemble dimensions at the bedside. For example, a physician might need to adjust dose downwards due to a patient's sensitivity to side effects. Or they might encounter patients who are insufficiently responsive at optimal doses—and hence candidates for higher dose. This "boundary information" around the critical points on the dose dimension will help bedside physicians make good decisions about the extent to which they can adjust the dose and still achieve a net therapeutic benefit.

The scientific reasons stem from the fact that informative clinical trials in the later phases of research depend upon the optimal intervention ensemble elements having been determined. Confirmatory trials are expensive endeavors, both in terms of the human and material resources consumed and the opportunity cost (Hey and Kimmelman 2013). Should a confirmatory trial "disconfirm" the clinical utility of an intervention ensemble, the disconfirmation is far more informative to the medical community if there are good grounds for believing that the drug was tested at or near the optimal dose, with the appropriate timing, in the appropriate diagnostic population, etc. If a negative confirmatory trial cannot plausibly eliminate the possibility that there are other viable—and as yet undiscovered—peaks in the landscape, then a very expensive human experiment has been used to explore only one small region of the ensemble space. This is not only inefficient, it is also excessively burdensome for patients and unsustainable for the research enterprise.

Finally, the policy reasons for locking down ensemble dimensions relate to the costs of uncertainty for health-care systems. As we described above, over and under-prescription carry significant health risks for the general population. Health-care systems can and do collect some of the boundary information necessary to avoid errors in prescription practices, but they are far less efficient at doing so than the research system. This is partly because health-care systems do not implement the kinds of practices that reduce the effects of bias. Also, health-care systems—unlike trial protocols—are not generally designed to aggregate and synthesize information across different treatment encounters. It is therefore in the interest of health-care systems (as well as the general patient population) to have a robust understanding of an intervention ensemble's risk/benefit landscape before it is taken into clinical practice.

11.4 Elucidating Ethical and Scientific Aspects of Early Phase Trials

All major research ethics policies instruct investigators to maintain a favorable risk/benefit balance. This entails minimizing risk for subjects, which means (a) minimizing the number of patients exposed to burdens in clinical research and (b) minimizing the total burden for each subject. Early phase researchers are therefore obligated to map the clinically-relevant regions of ensemble space using the fewest patients, with the least burden. Late phase researchers are obligated to test only intervention ensembles close to reasonably well-characterized optima in ensemble space.

In this section, we describe five nettlesome ethical and policy problems surrounding the ethical justification of risk in research, and how the concept of ensemble space can aid in addressing each.

11.4.1 Initiation of Early Phase Testing

When should clinical development of a new drug be initiated? Given some irreducible uncertainty surrounding first-in-human studies, it remains an open ethical and methodological question precisely when researchers know enough to begin clinical development with minimal risk to patient-subjects.

The ensemble concept can help to clarify this issue: Launch of clinical development often rests on pre-clinical evidence derived from animal testing. Just as early phase studies should map the intervention ensemble space for a new drug, animal studies should map the corresponding pre-clinical ensemble space. In other words, these animal studies can and should be used not merely to demonstrate that a drug has clinical activity (as is often emphasized), but also to search for necessary and sufficient conditions for unlocking clinical utility of a drug (Kimmelman et al. 2014).

While the pre-clinical space is never a perfect analog, it can nevertheless provide researchers with valuable inferential evidence about the contours and relevant dimensions of human intervention ensemble space.

For example, if timing of intervention application is likely to matter relative to disease course, pre-clinical researchers should strive to map, at least roughly, the optimal window of drug administration—much as they would for dose—with dose-response curves and pharmacokinetic studies. If co-interventions are likely to be necessary to achieve therapeutic activity in humans, pre-clinical studies should map the dimension of co-intervention application—studying variables like co-intervention dose and interactivity with the study drug.

Thus, initiation of clinical development should only begin once early phase researchers have justified estimates for (1) the relevant dimensions that need to be explored, (2) the range of each dimension parameter, and (3) a target optimum. To put it simply: clinical development should begin only when researchers can construct and justify a complete representation of ensemble space. Early phase studies should therefore work in concert with pre-clinical studies to narrow the area of the ensemble space explored in clinical trials, thereby minimizing patient exposure to toxic and/or inactive intervention ensembles.

11.4.2 Subject Selection in Early Phase Research

Another recurring ethical issue emerges in the context of neurodegenerative diseases, such as Alzheimer's disease (AD) or Amyotrophic Lateral Sclerosis (ALS). Many commentators debate whether early phase trials should enroll patients who have recent disease onset or patients whose disease is advanced. The former patients have the most to gain from participation, in the event that the experimental intervention is found to be effective. However, they also have the most to lose in the event that the experimental intervention is found to be either inert or harmful—and, unfortunately, this outcome far more probable, given low base rates of success in clinical translation for neurological disease.

Enrolling patients with advanced disease is typically considered the safer option. Although these patients have less to gain from participation in a positive trial, they also have less to lose, since their quality of life will have already deteriorated. This makes the consequences of a negative trial potentially less disastrous. However, only enrolling advanced-disease patients can also make a trial less informative. For example, if the experimental intervention is hypothesized to have some prophylactic effect, this will not be easily observable in patients with advanced disease.

This dilemma of subject selection can be understood as a trade-off between two kinds of decision-theoretic strategies: One of these is a maximax or "innovative care" strategy, which aims to maximize the benefit to research participants. This approach begins by exploring the region of the estimated peak on the landscape. Although, by definition, this peak is not yet known, the innovative care approach draws on whatever evidence is available (whether pre-clinical or derived from analogous interventions) to project where this peak may be, and then initiates the trial there—i.e., using the particular dose, schedule, timing, co-interventions, etc. that are projected to produce maximum benefit.

The other approach is a maximin, or "risk-escalation" strategy, which aims to minimize the greatest possible harms, both in terms of harm to participants and harm to the integrity of the research enterprise. This approach begins by exploring the neighborhood of the only landscape region that is known at the outset of testing—the region of low-risk near the origin, which we call the "base region." This would mean, for example, giving the first cohort of patients a sub-therapeutic dose with the least invasive mode of administration. The strategy would then escalate the risks over a series of cohorts—systematically searching the landscape for the critical points.

In terms of subject selection, enrolling recent onset disease patients in a high-risk, high-reward trial is an innovative care approach; enrolling advanced disease patients in a lower-risk, lower-reward trial is a risk-escalation approach. The dilemma for early phase researchers is which of these two approaches to use. Although there are circumstances under which the innovative approach is necessary, the risk-escalation approach is generally the better of the two. Ensemble space can help to demonstrate why this is so.

Figure 11.2 illustrates the best and worst-case scenarios for the innovative care approach. In the best-case, the peak of benefit signal is exactly where it was projected to be. This is a boon not only for the patient-subjects enrolled in the trial, who may experience a Lazarus-like response, but also for the sponsors, who will save money from an expedient translation effort; the researchers, who may reap the career benefits of a bold scientific discovery; as well as the health-care system, which will have more rapid access to the new intervention ensemble. However, we should note that even in this best case, there is little information gained about the lower and upper-bound critical points.

In a worst-case scenario for the innovative care approach, the peak is not where it was projected to be. Patient-subjects in the first cohort will have been exposed to considerable risks—perhaps even killed—and again, little information will have been gained about the critical points on the risk/benefit landscape. This means that the trial has not delivered the kind of information trialists need to protect subsequent patient cohorts. This is because the researchers do not have a clear idea of the shape of the landscape—for example, how much they need to lower dose to land in a safer, active region of the landscape. Particularly in high-profile areas of research, this kind of failure can lead researchers and funders to shift their focus to other areas that are perceived to be less risky or failure-prone (Cohen 2002; Holden 2009; Redmond 2002; Wilson 2009).

By contrast, Fig. 11.3 illustrates the best and worst-case for the risk-escalation approach. Because researchers often know that certain dimensions scale with risk—that, for example, increasing dose also increases risk—researchers will have a good sense of how to extend their exploration beyond the base over several iterations. In the best case, the peak of benefit signal is close to the base region, and can therefore be discovered after enrolling only a few patient cohorts. Although this is likely

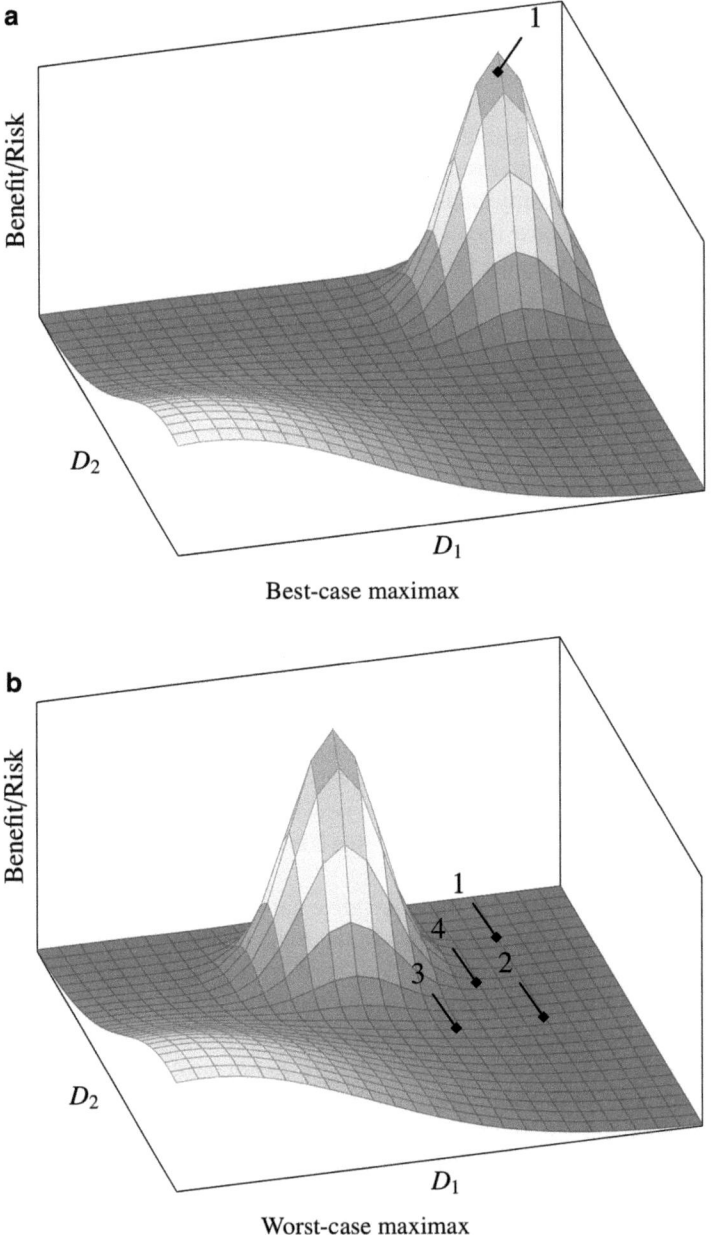

Fig. 11.2 Best- and worst-case outcomes for the innovative care strategy. (**a**) The peak of benefit signal corresponds almost exactly to the estimates from preclinical data. Only one trial is needed to identify a promising ensemble, which can then be advanced into later phase trials. (**b**) The peak does not correspond to the preclinical estimates, and the deaths or toxicities in each negative trial do not provide informative evidence about how to adjust ensemble parameters. Each subsequent trial is then an arbitrary and risky guess in the neighborhood of the original preclinical estimate

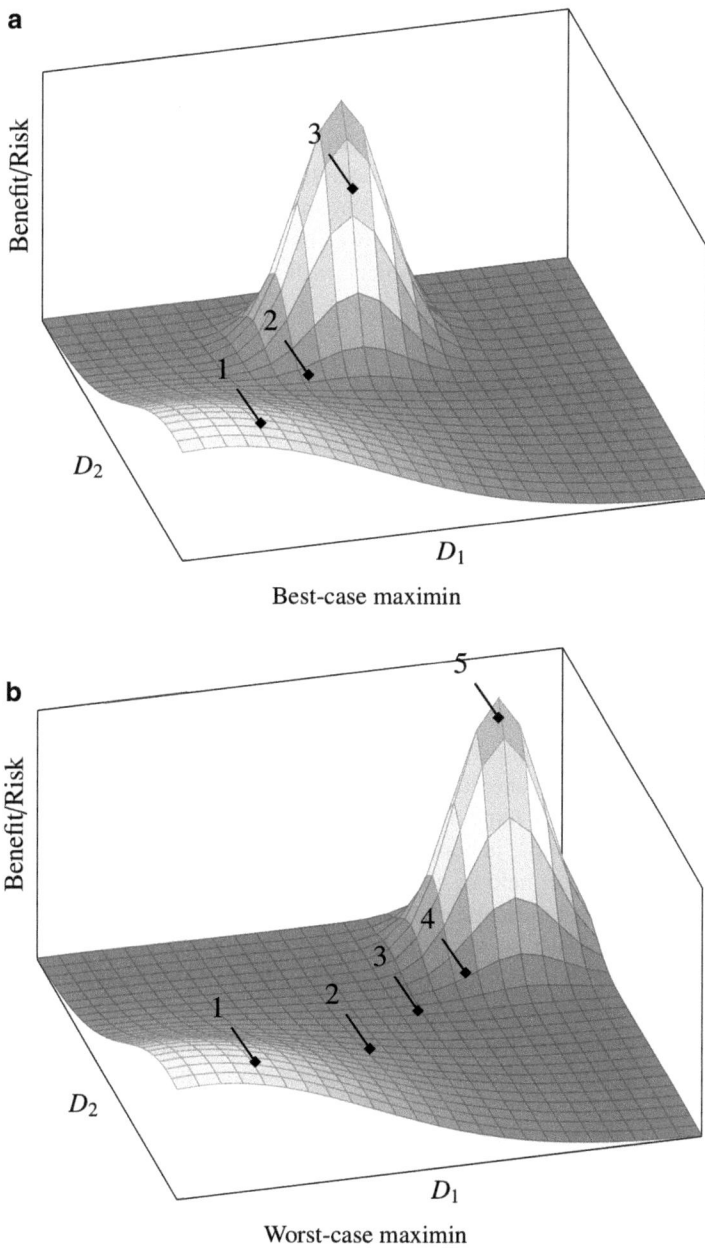

Fig. 11.3 Best- and worst-case outcomes for the risk-escalation strategy. (**a**) The peak of benefit signal is discovered after few trials, exposing patients to minimal risks. (**b**) The peak is deep into the high risk end of the landscape and many trials, of increasing risk, are needed to discover it

more expensive and time-consuming than the innovative care approach, risks will have been minimized throughout, and at least the lower-bound critical point will be determined.

In the worst-case scenario, the peak is far out into the landscape, and will only be discovered after several cohorts. This requires a sustained investment of human and material resources, which may tax the research stakeholders considerably. However, even for this worst-case, the risk-escalation approach ensures (a) that the peak will eventually be discovered; (b) that the lower-bound critical point will be identified efficiently; and (c) risks to patient–subjects will have been minimized throughout.

11.4.3 Initiation of Randomized Trials

Few questions have generated as much controversy in trial ethics as that of when it is ethical to randomly allocate patients to a new treatment. The principle of clinical equipoise demands that for an RCT to be ethical, there must exist a state of genuine uncertainty amongst the expert medical community concerning the relative therapeutic merits of each arm of the trial (Freedman 1987). In other words, all the arms of an RCT must be consistent with competent care, and the trial is conducted in order to resolve the question of which is the best.

In most instances, the control arm of an RCT will be the standard of care and the experimental arm is the novel, as-yet unproven, therapeutic ensemble. The burden of establishing clinical equipoise thus falls on exploratory stages of research—these early phase trials must generate evidence sufficient to justify the state of genuine uncertainty. However, it remains controversial just when our understanding of a new intervention ensemble is mature enough to justify randomized trials.

Ensemble space can help to address this controversy by illuminating how the "known" risk/benefit landscape evolves over the course of the discovery stage. At the outset of clinical testing, two things can be reliably inferred about a landscape: (1) the location of the base region; (2) the location of our best estimates for the various critical points, as derived from pre-clinical evidence. However, these pre-clinical estimates are just that: they are rough estimates that can be used to guide the direction of exploration, but for most disease domains, they cannot be assumed to be correct.

In Fig. 11.4, we illustrate how understanding of the landscape evolves over time (following a risk-escalation strategy). As we show, RCTs should be only initiated once four conditions have been satisfied: (1) the necessary and sufficient values of the intervention ensemble dimensions have been established; (2) the lower-bound critical point is known; (3) a peak of clinical interest for relevant dimensions has been discovered; and (4) clinical activity at the peak is believed to be competitive with standard of care.

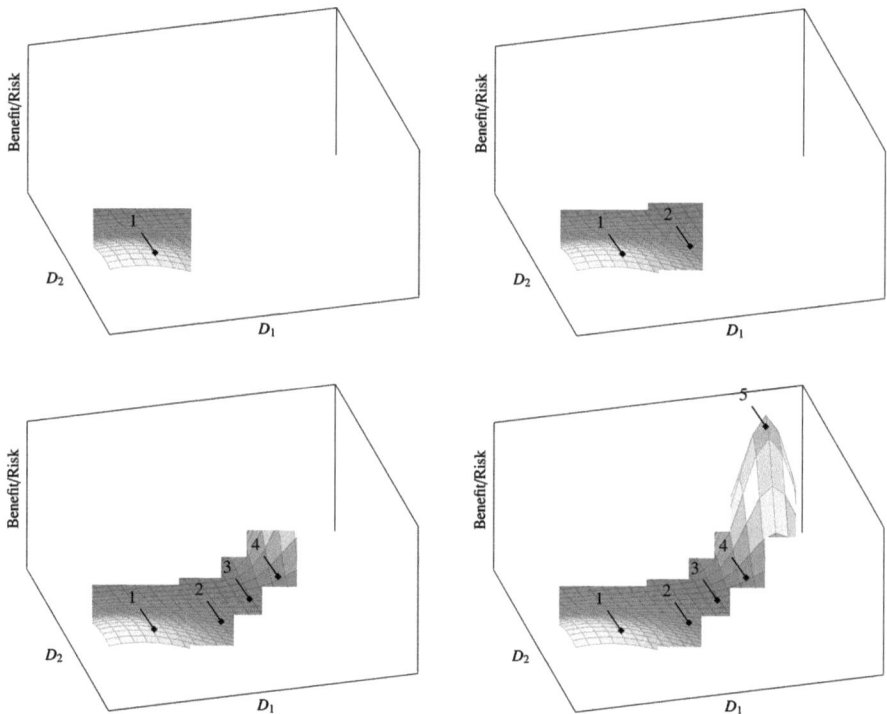

Fig. 11.4 Evolution of knowledge over time. The lower boundary (discovered at point 4) and peak (discovered at point 5) together are sufficient information to initiate randomized trials

11.4.4 Ethical Justification of Risk in Early Phase Research

Various commentators contend that early phase studies can—and should—have a therapeutic orientation (Markman 1986). Others have argued that early phase studies should not be viewed as therapeutic endeavors (Ross 2006). How this debate is resolved shapes the design of trials, as well as the informed consent process. For example, accelerated dose escalation strategies and adaptive randomization aim at maximizing the number of patients receiving "therapeutic" levels of a drug in early phase studies; such methodologies are to be favored if one can defend a therapeutic orientation for early phase studies. On the other hand, if application of a drug in early phase studies lacks therapeutic justification, design strategies that minimize patient exposure and risk by using smaller patient cohorts or safer intervention strategies are to be favored.

The concept of ensemble space helps cement the case against phase studies having a therapeutic orientation (Anderson and Kimmelman 2010). As the landscape representation makes explicit, the vast majority of positions in ensemble space are at best benign and at worst toxic. Research is initiated on the hypothesis that there is a peak somewhere in the landscape, but the location of this peak is not yet known. This entails that, until the region of the landscape containing therapeutic activity has been clarified, there should be no presumption of therapeutic benefit for volunteers in early phase studies.[5] Contrast this with decision-making in late phase testing. If early phase studies were performed properly, researchers should have confidence that they have identified all necessary and sufficient conditions to unlock the clinical utility of an intervention ensemble. Uncertainties in this case are of a narrow type: namely, the risk/benefit associated with the intervention ensemble, rather than the identity and values of the intervention ensemble components.

However, early-phase studies following a risk-escalation strategy can presume benefits in terms of scientific knowledge. Each study will contribute to the understanding of the overall ensemble space.

In sum, when understanding is mature, then potential direct benefits can justify exposing volunteers to risks. Until that time, knowledge value must justify the risks. This emphasizes the need to situate a trial's objectives within the context of scientific understanding. What question is it addressing? Which dimensions still need to be locked down before randomized trials can be justified?

11.4.5 Reporting Trials

Up to now, policy debates on reporting of clinical research have centered on confirmatory trials. For instance, FDA policy does not obligate prospective registration of phase 1 studies (USFDA 2007), and reporting guidelines for early phase studies are scarce.[6] The emphasis on confirmatory trial reporting reflects a misconception that early phase studies do not connect with the realm of clinical practice.

Ensemble space helps to elucidate the direct relationship between early phase studies and clinical practice. We noted that medicine is practiced not by applying drugs, but by applying intervention ensembles that contain drugs. Caregivers often adjust intervention ensemble dimensions that have been validated in confirmatory trials because of individual patient characteristics—co-morbidities, sensitivities, timing of presentation, etc. This process of adjustment draws on knowledge of the landscape and critical points within an ensemble space. Inability to access this information can lead to delivery of ineffective care.

---

[5] We note that, at the outset of testing, researchers may have hypotheses about the identity and values of all intervention ensembles. It is conceivable that, if these hypotheses are grounded in good evidence, one can project therapeutic benefit for patients even during exploratory stages of research. We think such conditions may hold, under some circumstances, in phase 1 pediatric oncology research, where dosing and other intervention ensemble components are heavily informed by adult data and relatively mature knowledge of pediatric physiology.

[6] See the EQUATOR website: www.equator-network.org.

11.5 Potential Problems and Limitations

Thus far, we have sketched how the concept of ensemble space helps articulate criteria for pre-clinical study design, trial initiation, trial design, and reporting. However, the concept also simplifies the process of intervention development, and there are important ways in which the path to developing a drug resists representation as a multi-dimensional ensemble space.

For example, not all of the relevant dimensions of a new intervention ensemble are known in advance of clinical development. Some dimensions only come into view after extensive exploration. A simple way this might occur is if, after a series of early phase studies, researchers discover that certain previously uncontemplated co-interventions are necessary to unlock the clinical utility of a new drug.

A more subtle way dimensions can be added during clinical development is when researchers discover over the course of testing that some population characteristic modulates the clinical utility of a drug. Consider the way cancer drug developers often apply a new drug to many different malignancies early in development: often, such exploratory research activities will discover that a drug has activity against some malignancies but not others. It might take many trials—and further basic research—to discover the particular property that explains the patterns of activity observed in clinical development. In such cases, a key intervention ensemble dimension only becomes apparent after drug development is well underway.

This problem helps drive home the point that clinical trials, in addition to providing information about clinical utility, also supply information about causal processes underlying a drug's clinical utility. Indeed, some cancer drug developers are experimenting with new trial designs aimed at accelerating the discovery of such dimensions. For instance, in "basket trials," researchers apply a new drug in an unselected patient population with the aim of identifying "exceptional responders." Then, next generation sequencing of tumor tissue is used to determine whether there are certain molecular signatures shared by exceptional responders that predict efficacy (Willyard 2013). Such signatures, if discovered, might be represented as new intervention ensemble dimensions that can be systematically mapped.

Some ensemble dimensions may also require large populations to map, and these will be inaccessible until late phase trials (or beyond). Some adverse event rates, for example, may be undetectable until hundreds or thousands of patients–subjects have been exposed. Although this does not make the concept of ensemble space useless for mapping safety events, it does reduce its utility for early phase research, since an important clinical dimension cannot be explored until later phases.

Finally, we should acknowledge an interpretive challenge for the concept of ensemble space. Specifically: is the representation a literal map of the population-level risk/benefit landscape, which serves as both an ethical guide for researchers, as well as a scientific reference for clinicians? Or is it merely a methodological heuristic, which may be useful for formalizing and understanding decision-making in the practice of research, but does not truly capture any of the scientific properties of the intervention ensemble? In other words, who should be using the ensemble space to inform their practices?

At a minimum, we believe the concept of ensemble space has heuristic value. None of the arguments we presented above, illustrating how the concept can inform or resolve long-standing ethical issues in the early phases of research, depends upon a literal mapping. Indeed, the fundamental value of the tool lies in helping to structure thinking around risks in research—particularly in the early phases, where there is greater uncertainty about the properties of the experimental intervention. Even for domains with a very immature causal understanding, ensemble space could still be used for strategic or communicative reasons.

However, a literal interpretation of the ensemble space model is also not implausible. In domains with a mature causal understanding—where there are fewer dimensions of uncertainty and these scale smoothly with risk and benefit—it may be feasible to map the landscape and construct a working representation that can inform clinical practice. One can then explore the efficiency with which a drug development program mapped ensemble space.

11.6 Conclusion

Ensemble space provides traction on debates that have dogged investigators and research ethics committees for decades. It also inspires important new questions about research efficiency and risk minimization. Consider that exploration of ensemble space typically takes place across many different studies: in cancer, phase 1 studies might map the dose dimension, and phase 2 studies might map diagnostic indications. What kinds of trial designs or trial programs maximize the moral efficiency with which optima are discovered? What kinds of trial coordination practices maximize this moral efficiency? When all is said and done in early phase research, what fraction of patients receive drug near the proximity of optima, and what fraction receive drug outside the boundaries of clinical utility? Are certain dimensions explored more efficiently than others? How frequently do drugs fail clinical development because of a lack of clarity on known intervention ensemble dimensions?

The true test of the concept's value will be whether it can improve the moral economy of drug development by minimizing patient burden while maximizing the yield, accuracy, and precision of evidence for further drug development and clinical practice.

References

Anderson, J.A., and J. Kimmelman. 2010. Extending clinical equipoise to phase 1 trials involving patients: Unresolved problems. *Kennedy Institute of Ethics Journal* 20(1): 75–98.

Cohen, J. 2002. The confusing mix of hype and hope. *Science* 295(5557): 1026.

Freedman, B. 1987. Equipoise and the ethics of clinical research. *New England Journal of Medicine* 317(3): 141–145.

Hey, S.P., and J. Kimmelman. 2013. Ethics, error, and initial trials of efficacy. *Science Translational Medicine* 5: 184fs16.

Hey, S.P., and J. Kimmelman. 2014. The risk-escalation model: A principled design strategy for early-phase trials. *Kennedy Institute of Ethics Journal* 24(2): 121–139.

Holden, C. 2009. Fetal cells again? *Science* 326(5951): 358–359.

Kimmelman, J. 2012. A theoretical framework for early human studies: Uncertainty, intervention ensembles, and boundaries. *Trials* 13: 173.

Kimmelman, J., J.S. Mogil, and U. Dirnagl. 2014. Distinguishing between exploratory and confirmatory preclinical research will improve translation. *PLoS Biology* 12(5): e1001863.

Markman, M. 1986. Opinion: The ethical dilemma of phase i clinical trials. *CA: A Cancer Journal for Clinicians* 36(6): 367–369.

Piantadosi, S. 2013. *Clinical trials: A methodologic perspective*. New York: Wiley.

Redmond, D.E. 2002. Book review: Cellular replacement therapy for parkinson's disease—where we are today? *The Neuroscientist* 8(5): 457–488.

Ross, L. 2006. Phase I, research and the meaning of direct benefit. *Journal of Pediatrics* 149(1): S20–S24.

US Food and Drug Administration (USFDA). 2007. *Food and drug administration amendments act (fdaaa) of 2007*.

Willyard, C. 2013. "Basket Studies" will hold intricate data for cancer drug approvals. *Nature Medicine* 19(6): 655.

Wilson, J.M. 2009. A history lesson for stem cells. *Science* 324(5928): 727–728.

Woods, D., E. Thomas, J. Holl, S. Altman, and T. Brennan. 2005. Adverse events and preventable adverse events in children. *Pediatrics* 115(1): 155–160.

Chapter 12
Rethinking Risk–Benefit Evaluations in Biomedical Research

Annette Rid

Abstract One of the key ethical questions in biomedical research is how we should evaluate the risks and potential benefits of research studies. This essay suggests that the current framework for risk–benefit evaluations is not comprehensive and arguably places too much emphasis on informed consent as a condition of acceptable net risk to participants. Instead, it suggests that the scientific and social value of biomedical research is likely fundamental to the acceptability of exposing participants to net research risks. The essay offers a vision for a comprehensive framework for risk–benefit evaluations that revolves around the relation between net risk to participants and the social value of the research, while giving participants' informed consent an important role, and identifies questions for future research.

12.1 Background

One of the fundamental ethical concerns about biomedical research is that it poses risks to some individuals primarily for the benefit of others. Concerns about exploitation, vulnerability or undue inducement in research would gain significantly less traction if participants were exposed to risks entirely for their own benefit, rather than for the benefit of science or medical progress. A key question in research ethics therefore is: under what conditions are the risks to individual participants acceptable in light of the potential social benefits of the research? An answer is essential to protect participants from being exposed to excessive research risks, while ensuring that valuable research is not delayed or stifled out of unwarranted concerns about risk.

Despite their fundamental importance, frameworks for risk-benefit evaluations of biomedical research biomedical research remain surprisingly vague. Thirty-five years ago, the Belmont Report acknowledged the "metaphorical" nature of most

---

A. Rid (✉)
Department of Social Science, Health & Medicine, King's College London, London WC2R 2LS, UK
e-mail: annette.rid@kcl.ac.uk

approaches to evaluating whether the risks and potential benefits of research are "balanced" or produce "a favorable ratio" (NCPHSBBR 1979). More than 20 years later, the U.S. National Bioethics Advisory Commission observed that research regulations and guidelines uniformly mandate evaluation of research risks and benefits, but fail to explain how this is to be done (NBAC 2001). Today, several general frameworks for risk–benefit evaluations exist (London 2006; Rajczi 2004; Rid and Wendler 2010, 2011; Weijer 2000; Wendler and Miller 2007), and there is a growing literature on specific concepts incorporated into these frameworks (e.g. the "minimal risk" threshold in research without informed consent (Ackerman 1980; Freedman et al. 1993; Kopelman 2004; McCormick 1976; Resnik 2005; Rid et al. 2010; Ross and Nelson 2006; Wendler 2005; Westra et al. 2011; Rid 2014a), "equipoise" in randomized-controlled clinical trials (Freedman 1987; Miller and Brody 2003; Miller and Weijer 2007; van der Graaf and van Delden 2011)). However, as I suggest in this essay, current thinking about risk–benefit evaluations in research is still dominated by the idea that informed consent is the key condition of acceptable risk to participants. I submit that this idea is not compatible with our intuitions about the role of consent in the research context. Instead, the social value of biomedical research and particular research studies is likely fundamental to ensure the acceptability of exposing participants to research risks. I sketch the potential implications of this view for risk–benefit evaluations and the ethical oversight of biomedical research, while identifying key questions for future conceptual and normative analysis.

12.2 The Need for a Better Framework for Risk–Benefit Evaluations

To be ethically appropriate, biomedical research must satisfy a number of requirements, including the requirement of a reasonable risk–benefit ratio (Emanuel et al. 2000). Most commentators, and essentially all research guidelines and regulations, agree that it is justifiable to expose research participants to some level of risk for the benefit of others when two conditions are met. First, the risks to participants should be proportionate or reasonable in relation to the potential clinical benefits for them or the potential social benefits of the research. This implies that the risks of research interventions should be either justified by the potential clinical benefits for participants or, if this is not the case and an intervention poses "net risks", the risks should be justified by the scientific or social value of the research. Second, when participants cannot give their own informed consent (e.g. children, psychiatric patients), or they do not give informed consent for reasons of feasibility (e.g. waivers of consent for secondary uses of existing data) or for methodological reasons (e.g. research involving deception), any net risks to participant should be minimal (CIOMS 2002; CoE 2005; Emanuel et al. 2000; WMA 2013).[1] These points of consensus are not

---

[1] Some guidelines and regulations allow a "minor increase" over minimal risk in certain types of pediatric research (e.g. NBAC 2001; WMA 2013), though this risk category is not widely used.

only reflected in most guidelines and regulations, but also incorporated in most ethical frameworks for risk–benefit evaluations in research (London 2006; Rid and Wendler 2011; Weijer 2000; Wendler and Miller 2007). In addition, many guidelines and regulations calibrate the level of ethical oversight to the risks posed to participants (so-called "risk-adapted" systems of oversight). While they generally require prospective research ethics committee (REC) review for biomedical research studies, many guidelines and regulations allow expedited or no REC review for certain types of minimal risk research.

Although important, these points of consensus do not provide a sufficiently concrete and robust framework for risk–benefit evaluations and risk-adapted ethical oversight of biomedical research. Moreover, the academic literature in this area largely focuses on questions that pertain to no more than a relatively small number of research studies. Most of the literature seems to cluster around two questions. The first is the question of how to set the minimal net risk threshold in studies with participants who cannot or do not give informed consent—in particular, how to define and implement this threshold (Ackerman 1980; Freedman et al. 1993; Kopelman 2004; McCormick 1976; Resnik 2005; Rid et al. 2010; Ross and Nelson 2006; Wendler 2005; Westra et al. 2011; Rid 2014a). The second question regards the evaluation of the risk–benefit profile of study interventions with a prospect of clinical benefit (e.g. investigational drugs). Here, the focus is largely on the role of "equipoise"—the uncertainty or disagreement about the relative merits of the study intervention and any established treatment(s)—and exactly what equipoise is; whether it is a necessary or sufficient condition for acceptable risks to participants; or whether it is no condition at all (Freedman 1987; Miller and Brody 2003; Miller and Weijer 2007; van der Graaf and van Delden 2011).

While these discussions are valuable, a comprehensive and sufficiently detailed framework for evaluating research risks and potential benefits is still lacking. In particular, there is not enough guidance on how to distinguish different levels of risk to participants, as well as how to determine that the conditions of acceptable research risk are met at different risk levels. For example, for research involving greater than minimal risks, current guidance merely requires that the risks be reasonable in relation to the scientific or social value of the research and the informed consent of participants be obtained (CIOMS 2002; CoE 2005; Emanuel et al. 2000; WMA 2013). Arguably, however, it would be useful to distinguish two or three subcategories of risk within the broad category of "greater than minimal" risk and specify what amount or type of scientific or social value is required to justify the risks to participants at each level; what level of scrutiny is required in the informed consent process—and in safety management—as the risks to participants increase; and what kind of ethical review and scrutiny is adequate in research involving significant risks. For example, a more transparent and inclusive ethical review than the standard REC appraisal may be needed in high-risk research to ensure that judgments about acceptable risk are justified from a procedural perspective. The lack of sufficiently detailed guidance on how to distinguish different risk levels, as well as conditions of acceptable risk at each level, also makes it difficult to develop robust risk-adapted forms of ethical oversight.

This situation poses two important practical challenges. The first challenge is the documented variation and inconsistency of risk judgments between RECs. Several studies and anecdotal evidence suggest that RECs differ significantly regarding which risks to participants they deem acceptable (Lenk et al. 2004; Shah et al. 2004; van Luijn et al. 2002; van Luijn et al. 2006). This raises concern that participants are not always adequately protected from excessive research risks, while valuable research may sometimes be rejected out of unwarranted concerns about risk. The second challenge is the administrative burden associated with the current system of ethical oversight, which—as many argue—delays or even stifles valuable research for overall marginal gains in subject protection. Regulators are currently developing more risk-adapted mechanisms of oversight to address this concern; however, several fundamental questions about these efforts remain unanswered (Rid 2014b). In my view, a comprehensive and detailed ethical framework for risk–benefit evaluations in biomedical research is needed to address both of these challenges.

12.3 The Fundamental Question: What Justifies Exposing Participants to Risk?

Any comprehensive and detailed framework needs to rest on solid normative grounds regarding what *fundamentally* justifies exposing study participants to net research risks (i.e., the risks of a research intervention are not, or not entirely, offset by its potential clinical benefits for participants). Tellingly, commentators have addressed this question primarily in the context of research that involves participants who cannot give their own informed consent, especially children (Ackerman 1980; Brock 1994; Freedman et al. 1993; Ross 1998; Wendler 2010). For example, some commentators have argued that it is justified to expose children to some net research risks because research participation offers them educational benefits (Ackerman 1980) or the opportunity to contribute to a valuable project that makes the children's lives go better overall (Wendler 2010). There is no comparable literature on what justifies exposing competent participants to net research risks; this suggests that net risks are not—or significantly less—concerning when informed consent can be obtained. Indeed, the current framework for risk–benefit evaluations puts great emphasis on informed consent as a condition of acceptable research risk. Most guidelines and regulations allow no more than minimal net risks in research without informed consent. By contrast, there is no explicit upper risk limit when informed consent is obtained, provided the net risks to participants are reasonable in relation to the scientific or social value of the research (CIOMS 2002; CoE 2005; Emanuel et al. 2000; WMA 2013).[2]

---

[2] Some commentators have recently explored upper risk limits in research with competent consenting participants (London 2006; Miller and Joffe 2009; Resnik 2012; Rid and Wendler 2011). However, to my knowledge these limits have not been endorsed by current guidelines or regulations.

Consent clearly has an important role to play when evaluating research risks. After all, most research interventions involve some level of intrusion into participants' privacy or bodily domain that generally requires permission. But, contrary to most current thinking, I do not believe that consent fundamentally justifies exposing participants to net research risks. If consent was fundamental, it would be either a sufficient or a necessary condition for acceptable research risk. Yet I would posit that most people do not think it permissible for an investigator to put study participants at a 50 % risk of death purely for research purposes, even if the investigator obtains valid informed consent and the study has tremendous public health value (e.g. by evaluating a promising strategy for curing HIV/AIDS). Conversely, nearly everyone would arguably agree that it may be permissible to conduct important research without informed consent (e.g. research on childhood vaccinations) when the risks are sufficiently low. This suggests that consent, while important, is neither a sufficient nor a necessary condition for acceptable risk in research.

To properly ground these intuitions, we need to analyze what fundamentally justifies exposing study participants to research risks and the role of consent in this context—in research both without and with informed consent. We also need to probe what limits net research risks independent of informed consent (e.g. the uncertainty of societal benefits from any given study, the need to maintain public trust in research, considerations related to justice or the professional integrity of investigators (London 2006; Miller and Joffe 2009; Resnik 2012; Rid and Wendler 2011) and how we can delineate upper thresholds of acceptable risk). In addition, as some commentators suggest (Bromwich and Rid 2014; Miller and Wertheimer 2011; Sreenivasan 2003), further work is needed on whether informed consent can and should be adapted to the level of risk posed to participants. Yet although further analysis is clearly necessary, it already seems clear that the current framework for risk–benefit evaluations puts too much emphasis on informed consent as a condition of acceptable research risk. We thus need to develop a framework for risk–benefit evaluations that is not only comprehensive and detailed, as suggested above, but also specifies the proper role of consent.

12.4 Rethinking Risk–Benefit Evaluations: The Importance of the Social Value of Research

Biomedical research is a social pursuit that aims to advance our understanding of human health and disease and improve the diagnosis, treatment, and prevention of disease in future patients. A plausible candidate for what fundamentally justifies exposing study participants to net research risks therefore is the scientific or social value of the research. If a study has no scientific or social value, it is not clear what makes it acceptable for investigators to expose participants to net risks. For example, a study of a "me-too" drug that is so similar to already approved treatments in terms of side effects, route of administration, cost and so on, has no obvious scientific or social value. This suggests that a comprehensive and detailed framework for risk–benefit evaluations might fundamentally revolve around the relationship between net risks to participants and the scientific or social value of the research, with greater value necessary to justify higher net risks—up to an absolute upper limit of acceptable net risk in research of tremendous social value.

As we have seen above, consent is clearly important as well. So what role might it play in a framework that fundamentally revolves around net risks and the scientific or social value of the research? A plausible starting point is that consent would be a necessary, but not a sufficient condition for exposing participants to *substantial* net risks of harm. Within the normatively grounded absolute limits of acceptable research risk, moderate or high net risks to participants would be acceptable only when the research has proportionate scientific or social value and the consent process is rigorous enough to allow them to effectively protect their rights and interests. Low net risks would be acceptable if the research has sufficient scientific or social value, even if the consent process is not or cannot be as stringent (e.g. a written agreement to participate is not obtained), or informed consent cannot be obtained at all. This approach would capture our intuitions both that informed consent is not sufficient to justify exposing participants to net risks (especially substantial net risks), and that informed consent is not always necessary for research involving low net risks.

Based on these considerations, the framework might divide the spectrum of acceptable research risk—from essentially no net risk to the independently determined absolute upper limit—into several levels of acceptable net risk. Each level would pair a given ratio of net risk and scientific or social value with appropriate safeguards and protections for participants and mechanisms of ethical oversight. As the ratio of net risk and scientific or social value increases, so would the stringency of protections and oversight—for example regarding independent ethical review, the consent process, measures to reduce and manage risks, and perhaps other safeguards or constraints. For instance, in studies involving very high net risks (e.g., a first-in-human study of a novel gene transfer agent that is administered to the brain under general anesthesia), it might be adequate that the prospective ethical review process is transparent and conducted by a national panel of experts including a period of public comment. Conversely, in studies involving very low net risks (e.g. medical records research with de-identified patient data), it might be appropriate to abandon the model of prospective ethical review and require investigators to simply register their studies. A selection of studies might then be reviewed in retrospect.

Such a comprehensive framework for risk–benefit evaluations would have clear conceptual and practical advantages. Unlike current approaches, it would rest on a sound normative justification for exposing participants to net research risks, and it would capture our intuitions about the role of informed consent in the research context. Furthermore, by specifying several net risk levels instead of the customary two (i.e. minimal and greater than minimal risk), the framework would offer a graduated, nuanced approach to risk–benefit evaluations and ethical oversight. It should therefore be able to address the two practical challenges mentioned above (see "The need for a better framework for risk–benefit evaluations"). First, the framework would offer much more detailed guidance for RECs and others to evaluate the risks and potential benefits of research studies. Provided the framework is clear and broadly accepted by relevant stakeholders, and RECs and others are both trained and motivated to use it, the framework should help to reduce unjustified variation and inconsistency of judgments about acceptable research risk. This, in turn, should help to reduce the chances that study participants are not adequately protected or, conversely, that valuable research may be rejected out of unwarranted concerns about research risk. Second, the framework would offer a basic structure for developing more risk-adapted systems of ethical oversight and should thereby contribute to focusing oversight on those studies that warrant in-depth ethical scrutiny.

Needless to say, all of these ideas require further analysis and development as well as empirical testing. Although the scientific or social value of research is a plausible candidate for justifying net risks to participants, this needs to be examined in greater depth. For example, the philosopher Alan Wertheimer has recently argued that net risks are justified even when the research lacks (obvious) scientific or social value, provided that participants are adequately compensated for risks that they agree to assume (Wertheimer 2014). Wertheimer's argument is unlikely to be endorsed by many, as it assumes—against the common wisdom in research ethics—that payment can offset net risks to participants. However, it rightly presses the point that we currently lack a robust justification for why net risks should be acceptable only when the research has scientific or social value. Moreover, the concept of "scientific or social value" itself requires clarification before it can be incorporated into the envisioned framework. There currently is no systematic account of what makes research scientifically or socially valuable; what precisely the relationship between scientific and social value is; and what makes research sufficiently valuable such that it justifies exposing participants to greater net risks (Rid and Wendler 2011).

Furthermore, specifying a comprehensive and nuanced framework for risk–benefit evaluations and risk-adapted ethical oversight raises several additional questions. One set of questions concerns how we can better conceptualize and classify net risks to participants, given that the existing risk classifications do not comprehensively reflect these risks and likely use an insufficient number of risk levels (Rid 2014b). Another set of questions regards the specification of risk-adapted mechanisms of ethical oversight, informed consent procedures, safety monitoring processes, and perhaps other regulatory safeguards or requirements. In particular, it is notoriously difficult to determine what level of ethical oversight or scrutiny in the informed consent process is adequate at a given level of net risk. We need more clarity on the extent to which ethical oversight, consent procedures, and so on should be adapted to other considerations than risk (Bromwich and Rid 2014; Rid 2014b). The fact that research frequently exposes participants to net risks is not its only morally salient aspect, and certain "non-risk" considerations should arguably trigger enhanced ethical scrutiny (e.g. use of deception) or consent procedures (e.g. controversial study aims). Finally, careful empirical research is need to test whether the framework can be effectively implemented in practice, and whether it helps to reduce unwarranted variation and inconsistency of judgments about acceptable research risk. The framework should only be implemented on a larger scale when such research has showed promising results and requires careful monitoring when it is in place. My hope is that the present essay offers sufficient motivation to make this work worthwhile.

12.5 Conclusion

One of the key ethical questions in biomedical research is how we should evaluate the risks and potential benefits of research studies. However, the current framework for risk–benefit evaluations is not comprehensive and arguably places too much emphasis on informed consent as a condition of acceptable net risk to participants. In this essay, I have suggested that the scientific or social value of biomedical research is likely fundamental for ensuring the acceptability of exposing participants to net research risks. Future conceptual and normative work should probe this idea and examine the prospects of a comprehensive framework for risk–benefit evaluations that fundamentally revolves around the relation between net risk to participants and the social value of the research, while giving participants' informed consent an important role.

Acknowledgments Many thanks to Robert Goodin, Franklin Miller, David Wendler, and Alan Wertheimer for helpful comments on earlier versions of this essay. The research leading to these results has received funding from the Swiss National Science Foundation (PA0033-117505/2) and the People Programme (Marie Curie Actions) of the European Union's Seventh Framework Programme (FP7/2007-2013) under REA grant agreement n° 301816.

References

Ackerman, T.F. 1980. Moral duties of parents and nontherapeutic clinical research procedures involving children. *Bioethics Quarterly* 2(2): 94–111.
Brock, D.W. 1994. Ethical issues in exposing children to risks in research. In *Children as research subjects: Science, ethics, and law*, ed. M. Grodin and L. Glantz. New York: Oxford University Press.
Bromwich, D., and A. Rid. 2014. Can informed consent to research be adapted to risk? *Journal of Medical Ethics* 41(7): 521–528.
Council for International Organizations of Medical Sciences (CIOMS). 2002. *International ethical guidelines for biomedical research involving human subjects*. Geneva: CIOMS.
Council of Europe (CoE). 2005. *Additional protocol to the convention on human rights and biomedicine, concerning biomedical research*. Strasbourg: Council of Europe.
Emanuel, E.J., D. Wendler, and C. Grady. 2000. What makes clinical research ethical? *JAMA* 283(20): 2701–2711.
Freedman, B. 1987. Equipoise and the ethics of clinical research. *New England Journal of Medicine* 317(3): 141–145.
Freedman, B., A. Fuks, and C. Weijer. 1993. In loco parentis. Minimal risk as an ethical threshold for research upon children. *Hastings Center Report* 23(2): 13–19.

Kopelman, L.M. 2004. Minimal risk as an international ethical standard in research. *Journal of Medicine and Philosophy* 29(3): 351–378.

Lenk, C., K. Radenbach, M. Dahl, and C. Wiesemann. 2004. Non-therapeutic research with minors: How do chairpersons of German research ethics committees decide? *Journal of Medical Ethics* 30(1): 85–87.

London, A.J. 2006. Reasonable risks in clinical research: A critique and a proposal for the integrative approach. *Statistics in Medicine* 25(17): 2869–2885.

McCormick, R.A. 1976. Experimentation in children: Sharing in sociality. *Hastings Center Report* 6(6): 41–46.

Miller, F.G., and H. Brody. 2003. A critique of clinical equipoise. Therapeutic misconception in the ethics of clinical trials. *Hastings Center Report* 33(3): 19–28.

Miller, F.G., and S. Joffe. 2009. Limits to research risks. *Journal of Medical Ethics* 35(7): 445–449.

Miller, P.B., and C. Weijer. 2007. Equipoise and the duty of care in clinical research: A philosophical response to our critics. *Journal of Medicine and Philosophy* 32(2): 117–133.

Miller, F.G., and A. Wertheimer. 2011. The fair transaction model of informed consent: An alternative to autonomous authorization. *Kennedy Institute of Ethics Journal* 21: 201–218.

National Bioethics Advisory Commission (NBAC). 2001. *Ethical and policy issues in research involving human participants*. Report and recommendations of the National Bioethics Advisory Commission. Bethesda: U.S. Government Printing Office.

National Commission for the Protection of Human Subjects of Biomedical and Behavioral Research (NCPHSBBR). 1979. *The Belmont report. Ethical principles and guidelines for the protection of human subjects*. Washington, DC: U.S. Government Printing Office.

Rajczi, A. 2004. Making risk–benefit assessments of medical research protocols. *The Journal of Law, Medicine & Ethics* 32(2): 338–348.

Resnik, D.B. 2005. Eliminating the daily life risks standard from the definition of minimal risk. *Journal of Medical Ethics* 31(1): 35–38.

Resnik, D.B. 2012. Limits on risks for healthy volunteers in biomedical research. *Theoretical Medicine and Bioethics* 33(2): 137–149.

Rid, A. 2014a. Setting risk thresholds in research: Lessons from the debate about minimal risk. *Monash Bioethics Review* 32(1): 63–85.

Rid, A. 2014b. How should we regulate risk in biomedical research? An ethical analysis of recent policy proposals and initiatives. *Health Policy* 117(3): 409–420.

Rid, A., and D. Wendler. 2010. Risk–benefit assessment in medical research—critical review and open questions. *Law, Probability Risk* 9: 151–177.

Rid, A., and D. Wendler. 2011. A framework for risk–benefit evaluations in biomedical research. *Kennedy Institute of Ethics Journal* 21: 141–179.

Rid, A., E.J. Emanuel, and D. Wendler. 2010. Evaluating the risks of clinical research. *JAMA* 304: 1472–1479.

Ross, L.F. 1998. *Children, families, and health care decision making*. Oxford/New York: Clarendon.

Ross, L.F., and R.M. Nelson. 2006. Pediatric research and the federal minimal risk standard. *JAMA* 295: 759.

Shah, S., A. Whittle, B. Wilfond, G. Gensler, and D. Wendler. 2004. How do institutional review boards apply the federal risk and benefit standards for pediatric research? *JAMA* 291(4): 476–482.

Sreenivasan, G. 2003. Does informed consent to research require comprehension? *Lancet* 362: 2016–2018.

van der Graaf, R., and J.J. van Delden. 2011. Equipoise should be amended, not abandoned. *Clinical Trials* 8(4): 408–416.

Van Luijn, H.E., A.W. Musschenga, R.B. Keus, W.M. Robinson, and N.K. Aaronson. 2002. Assessment of the risk/benefit ratio of phase II cancer clinical trials by Institutional Review Board (IRB) members. *Annals of Oncology* 13(8): 1307–1313.

van Luijn, H.E., N.K. Aaronson, R.B. Keus, and A.W. Musschenga. 2006. The evaluation of the risks and benefits of phase II cancer clinical trials by institutional review board (IRB) members: A case study. *Journal of Medical Ethics* 32(3): 170–176.

Weijer, C. 2000. The ethical analysis of risk. *The Journal of Law, Medicine & Ethics* 28(4): 344–361.

Wendler, D. 2005. Protecting subjects who cannot give consent: Toward a better standard for "minimal" risks. *Hastings Center Report* 35(5): 37–43.

Wendler, D. 2010. *The ethics of pediatric research*. Oxford/New York: Oxford University Press.

Wendler, D., and F.G. Miller. 2007. Assessing research risks systematically: The net risks test. *Journal of Medical Ethics* 33(8): 481–486.

Wertheimer, A. 2014. The social value requirement reconsidered. *Bioethics* 29(5): 301–308.

Westra, A.E., J.M. Wit, R.N. Sukhai, and I.D. de Beaufort. 2011. How best to define the concept of minimal risk. *The Journal of Pediatrics* 159(3): 496–500.

World Medical Association (WMA). 2013. *Declaration of Helsinki: Ethical principles for medical research involving human subjects*. 64th WMA General Assembly. Fortaleza, Brazil. http://www.wma.net/en/30publications/10policies/b3. Accessed 15 Feb 2015.

Chapter 13
Towards an Alternative Account for Defining Acceptable Risk in Non-beneficial Paediatric Research

Sapfo Lignou

Abstract The purpose of this paper is to propose an alternative account by which the ethical threshold of acceptable risk in paediatric research can be assessed. Three popular interpretations of the minimal risk threshold and the problems they raise when applied in the research context are presented. First, the "risks of daily life" standard and the "routine examinations" standard are addressed. It is argued here that neither of them can provide a satisfactory morally justified framework within which risks during paediatric non-therapeutic research should be assessed. The alternative view of the "charitable participation" standard is then discussed and the argument advanced that despite its advantages, it generates unavoidable difficulties when considered in the context of medical research. Finally, the author argues that consideration of the risk to which parents are willing to expose their children in a vaccination programme in the case of an infectious disease, which does not constitute a significant threat to them, can facilitate the definition of this threshold. Although the proposed account shares some of its strong points with those already existing, it does not lead to inconsistencies when applied in research context.

13.1 Introduction

In current theory and practice, the predominant thesis is that subjects unable to give consent may be enrolled in research that is not designed to offer the prospect of direct benefit, only if several additional requirements are met. Arguably, the most important of these requirements is that participants should not be exposed to more than "minimal risk". Exceptionally, they may sustain only a "minor increase" over minimal risk, but only if the research conducted "is likely" to provide essential knowledge that is of "vital importance" for "the subjects' disorder or condition" (45 CFR §46.406).

S. Lignou (✉)
Division of Medicine, University College London, Gower Street, London WC1E 6BT, UK
e-mail: sapfo.lignou.09@ucl.ac.uk

However, although the minimal risk threshold is widely endorsed, its regulatory definition remains unclear. As a result, controversy exists on both the interpretation and the implementation of this rule (Glass and Binik 2008). This ambiguity translates into a lack of sufficient guidance for Institutional Review Boards (IRBs) and Research Ethics Committees (RECs), meaning that assessment is highly dependent on intuition (NBAC 2001, 71). This controversy is also evident in existing approaches to risk-benefit assessment for clinical research, none of which provides guidance on how to decide whether or not a given risk should be considered "more than minimal" (Rid and Wendler 2010).

Children[1] constitute the main population group in which research participation remains restricted (Tan and Koelch 2008; Tracey 2006) and for which group there is a high rate of off-label medication use (Korenman 2004). On the other hand, children are a vulnerable population, that is, they differ from competent participants in that they lack (adequate) autonomy to understand and assess research risks for themselves, and thus, consent for their enrolment in clinical research. Moreover, because of their vulnerability, they are susceptible to greater abuse; as it is well known, their welfare has been compromised in the past despite the existence of regulations for their protection (ACHRE 1995; Sharav 2004). Since it is only natural that medical research involves risks, ethical guidelines must balance the protection of children-participants with the significance of conducting research that can lead to medical improvement for this population group.

Non-beneficial clinical research constitutes an essential part of the process of clinical research. It is required for the evaluation of the safety and efficacy of medical interventions (Wendler and Glantz 2007); in other words, without it, therapeutic research could not be realised and thus medical improvements could not be achieved (Edwards and Wilson 2010).[2] The correct interpretation of a minor risk threshold is therefore crucial, since it serves to determine whether specific research is ethically designed, excessively restricted or fails to protect the integrity of its participants (Diekema and Stapleton 2006).

The purpose of this paper is to propose an alternative account by which the ethical threshold of acceptable risk in paediatric research can be assessed. Three popular interpretations of the minimal risk threshold and the problems they raise when applied in the research context are presented. First the "risks of daily life" standard

---

[1] The term child is used broadly here to refer to individuals below the age at which they can provide legal consent irrespectively of their capacity for consenting (following the definition of child in federal regulations on human research, which does not cite an age range: "persons who have not attained the legal age for consent to treatments or procedures involved in the research, under the applicable law of the jurisdiction in which the research will be conducted" (45 CFR 46.402(a); 21 CFR 50.3(o))). However, in the following paragraphs I argue that the concept of minimal risk should be age-adjusted.

[2] For instance, many Phase 1 studies, which are conducted to determine a safe dose of the drug under investigation, essentially offer no chance of medical benefit and pose at least some risks. Moreover, Phase 3 studies, which randomise subjects to a potential new treatment or existing standard treatment, typically include individual non-beneficial procedures, such as additional blood draws, to evaluate the drugs being tested.

and the "routine examinations" standard are addressed. It is argued here that neither of them can provide a satisfying morally justified framework within which risks during paediatric non-therapeutic research should be assessed. The fact that risks exist does not imply that the same level of risk should be morally acceptable in the research context. Two voluntary activities analogous with the research setting, chosen by parents for their social benefit, are discussed, from which IRBs/RECs may draw inspiration. It is argued that despite its advantages, the alternative view of the "charitable participation" standard generates inevitable difficulties when considered in the context of medical research. Finally, it is maintained here that consideration of the risks to which parents are willing to expose their children in a vaccination programme, in the case of an infectious disease, which does not constitute a significant threat to them, can facilitate the defining of this threshold. Although the proposed account shares some of its strong points with accounts that already existed, it does not lead to inconsistencies when applied in the research context.

13.2 Interpretations of the "Minimal Risk" Threshold

Many different accounts for the interpretation of the "minimal risk" threshold have been proposed, albeit without wider consensus. Those most frequently found in academic literature and national regulations are the "risks of daily life" standard, the "routine examinations" standard, and the "charitable participation" standard.

13.2.1 "Risks of Daily Life" Standard

According to the "risks of daily life" standard, the level of risks that children face in their daily lives constitutes the baseline for assessing the acceptability of the risks of the research in question.[3] This standard could be interpreted in relativistic (referring to the daily lives of the research subjects) or absolute terms (referring to the daily life of a normal, healthy person) (Resnik 2005).

A relativistic interpretation, as Kopelman has argued, could lead to an inequitable distribution of research risks, because the level of risks that various populations/people encounter in their daily lives can be very different. Thus, the adoption of this approach would lead to multiple inconsistencies in risk assessment by IRBs, since risks would be assessed differently, depending on where the research is conducted, and as a result ill children or children living in dangerous environments would be exposed to greater risks than those who are healthy or live in safe(r) environments. The widely used example of the Willowbrook hepatitis experiments has also illustrated that this interpretation can lead to dangerous consequences; according to the

---

[3] In some guidelines (Australia, Nepal and the U.S.) this definition is combined with the routine examinations standard (Wendler and Glantz 2007).

view of the researchers, the mentally disabled children that were infected with hepatitis were only exposed to minimal risk, because hepatitis was widespread in the institute (Resnik 2005).

Freedman, Fuks, and Weijer on the other hand maintain that a relativistic approach of minimal risk promotes flexibility and adaptability, and that minimal risk should be considered as a moral threshold that partially depends on community values: "the concept of risks of everyday life has normative as well as descriptive force, reflecting a level of risk that is not simply accepted but is deemed socially acceptable […] minimal risks are what we deem socially acceptable" (Freedman et al. 1993).

Despite the advantages of allowing IRBs greater flexibility when including local conditions and standards, we should consider that fairness and the integrity of the participants should not be violated—especially when they are vulnerable. Therefore, an absolute standard should apply to different localities. This position was endorsed by the Maryland appellate court and The National Bioethics Advisory Commission (Resnik 2005). However, even when adopting an absolute interpretation of daily life risks, this standard of assessing minimal risk remains problematic for several reasons.

First of all, daily risks cannot be easily identified and they are neither stable nor uniform (Resnik 2005); we cannot accurately quantify the degree of risk that children encounter daily (a general sense of this risk is not sufficient; researchers and IRB members should be guided by a clear and unambiguous definition and quantification of research risks) (Tracey 2006). Moreover, many of the daily risks to which children are subjected have unconscious acceptance (we don't actually know the probability and magnitude of risks of all our actions) or are involuntarily imposed by the activities that parents choose for their children (e.g., in some cases, children have to travel by car to go to school). Since paediatric research is an intentionally chosen risk-laden activity, the reasons for an analogous level of risk would not apply (Wendler and Glantz 2007).

Another very important difference between a daily life activity and a non-beneficial clinical trial is that many of the activities that parents allow for their children have the potential of benefit for them. Therefore, they may for instance allow their children to play sports not because these activities impose minor risks for their children's health (since the risk of bodily harm is high), but because they believe that these risks are outweighed by the benefits their child will gain (its physical and social development). However, the same risks may not be socially permissible for the purpose of producing generalisable medical knowledge that is not designed to offer personal benefit (Fisher et al. 2007; Marshall 2000).

Nevertheless, the following practical example is put forward as an objection to the above arguments: parents are morally justified to leave their child with a professional caregiver (a fact that often upsets or bothers children) so that they can have an enjoyable free evening. Although this is not an action all parents would follow, it is considered morally acceptable to put their child at this kind of risk[4] for their own

---

[4] As the authors note there is some risk in hiring a new person, particularly one unknown to the family, for the care of one's children. This is the reason why many parents feel a certain discomfort when they have to entrust the care of their children to someone else.

personal benefit (Glass and Ofenberg 1996). However, even this case is different from the case of non-beneficial research, since the benefit concerns the members of the family, the parents who took the decision and not the society in general (some other children who are unidentified).

It follows that the purpose of exposing children to daily life's risks is different from the purposes of exposing children to non-beneficial research. Therefore, the same level of risk considered acceptable in the former case may not be appropriate in the latter (Litton 2008; Wendler and Glantz 2007).

13.2.2 The "Routine Examinations" Standard

According to the "routine examinations" standard, research exposes its subjects to minimal risk when "the probability and magnitude of the harm or discomfort anticipated in research are not greater than those encountered during the performance of routine physical or psychological examinations or tests" (Resnik 2005). This thesis is strongly supported by many critics of the "risks of daily life" standard and was adopted by The Council for International Organizations and Medical Sciences (CIOMS) (Wendler and Glantz 2007) and Guideline 9 of Federal Regulations (Tracey 2006).

However, this interpretation presents some ambiguities as well; there is no agreement in the medical community on what constitutes routine examination (Resnik 2005), nor on the assessment of risks associated with various procedures, as illustrated by a survey by Janofsky and Starfield (1981) (see also Freedman et al. 1993). In addition, although both the US Food and Drug Administration (FDA) and Department of Health & Human Services (DHHS) have published lists of routine procedures to offer some guidance, it still remains difficult to assess the risks of new and complex research studies based on risks associated with medical or psychological test (Resnik 2005).[5]

Moreover, as in the case of the "risks of daily life" standard, problems associated with the subjective or absolute interpretation, also apply here. That is, it is not clear whether the routine examinations standard should be applied to the particular participant or to the healthy child (Weijer 2000). Those supporting the subjective interpretation argue that since such risks are familiar to the experience of the child, "such activities should be considered normal for these children" (Weijer 2000). However, this is problematic, mainly for two reasons. Firstly, the risks of "routine examinations" cannot be morally justified when these examinations do not benefit the recipients. As Wendler (2005) notes, "clinicians who expose children to routine examinations when the children cannot possibly benefit from the examinations are guilty of medical malpractice." The argument from analogy is then problematic, since non-beneficial research is not designed to benefit the participant.

---

[5]To assess new or complex studies, it is necessary to be able to appeal to a general concept of minimal risk, such as the "daily life risks" standard.

Secondly, following the subjective interpretation, ill children whose routine medical care involves risky examinations could be subject to greater risks than healthy children (Wendler and Glantz 2007). The National Bioethics Advisory Commission (NBAC) member Robert Turtle objected vigorously to this provision stating that "To do so, is to add to the potential burdens that result, directly or indirectly, from the child's illness" (Weijer 2000). In contrast to the aforementioned implications of the subjective interpretation, there are cases in which it is crucial that ill children be even more protected from the risks to which healthy children are routinely exposed (e.g., blood drawing procedures) since the same procedure can involve great risk of harm for them (e.g., children with haemophilia) (Fisher et al. 2007).

Hence, to respect fairness in the distribution of research risk, "minimal risk" should be defined as the level of risk that is posed by routine examinations for healthy children (Tracey 2006). This interpretation though, is also problematic, since the only invasive examination recommended for healthy children by the American Academy of Pediatrics, is a single heel stick at birth to screen for metabolic disorders (Wendler and Glantz 2007). Interpreted in a research context, it implies that no research, involving more than the level of risk than is posed by a heel stick, should be acceptable. This is a very restrictive standard, since its adoption would certainly reject as morally unjustified most of the non-therapeutic clinical research on children, including studies that seem intuitively acceptable, and thus will lead to undesirable consequences for their welfare (Tracey 2006). Finally, this interpretation appears even more problematic for countries in which children do not undergo routine medical examinations (Wendler and Glantz 2007).

Although the assessment of "minimal risk" by routine physical and psychological examinations presents some problems, we should note that it constitutes a more robust threshold than that provided by the "risks of daily life" standard, which is restricted by the lack of empirical information on everyday risks and the different normative judgments of socially-allowable risks for children (Kopelman 2004). In addition, the "routine examinations standard" enables the assessment of the probability and magnitude of harm or discomfort (posed by these examinations) according to the age of the child (Fisher et al. 2007, 7). However, as we have seen, the "routine examinations" standard also presents ambiguities when used as a baseline measure for assessing minimal risk.

13.2.3 The "Charitable Participation" Standard

An alternative standard to those discussed above is the "charitable participation" standard, proposed by Wendler. According to this standard, the risks of non-beneficial paediatric research should be assessed by the level of risks to which children may be exposed in appropriate charitable activities. He argues that this approach offers an objective standard, because of the similarities between charitable activities and non-beneficial research in many moral respects: they are both designed to help unidentified individuals, for whom there is only a possibility to benefit (Reynolds and Nelson 2008). Children's participation in charitable activities (such as planting crops, visiting a sick child, digging wells) under the appropriate safeguards—is widely considered morally justified, despite posing some risk to them. Similarly, it can be argued that it is morally justifiable to subject children to the risks of non-beneficial research when these risks are accepted by the society and their parents (Wendler and Glantz 2007).

However, as Wendler notes, there are several differences between non-beneficial research and charitable activities: first, the participation of children in charitable activities is active, in contrast to ordinary research procedures. Also, many charitable activities permit parents' or other adults' participation and involvement, whereas during research, medical interventions are imposed exclusively on children (Wendler and Glantz 2007).

Moreover, infants or toddlers do not typically participate in charitable activities, as they are unable to contribute. As Wendler and Grantz (2007) note, this may imply that in research settings, infants and toddlers should be subjected to a lower level of risk for the social benefit. Finally, they state that collection of empirical data is needed, based on the levels of risk to which it is right to expose children during their participation in charitable activities (Wendler and Glantz 2007).

According to this approach, it is clear that risks exist to which parents expose their children for the benefit of unidentified others. This exegetic element is lacking in both the "routine examinations" and the "risks of daily life" standards. However, the difficulties that Wendler notices in his approach are more serious than he acknowledges, rendering this standard problematic. First of all, he fails to point out that for many parents charitable activities constitute an opportunity for the moral development and socialisation of their child, as is the case with other daily activities (participation in sports, etc.). Although it can be argued that children's participation in non-therapeutic research is also considered by some parents as an opportunity for the moral development of their children, it seems easier to imagine a very young child understanding the virtue of altruism by participating in a charitable activity with other children, than by being subjected to the discomfort and anxiety of a medical intervention (however, this does not suggest that children are generally unwilling to participate in research in order to help other children).

In addition, Wendler and Grantz acknowledge that there is a difference between a charitable activity and a research process. They identify it as the lack of activeness of the child and the participation of its parent (Wendler and Glantz 2007). However, the most relevant fact is that medical procedures are usually unpleasant and thus may contain risks of psychological harm to the child (which, in addition, is detached from the parent), something very uncommon in charitable activities. Finally, although a toddler or an infant cannot contribute to a charitable activity in a research setting, their participation may be essential for the medical benefits of the equivalent age group, and it is important to consider how research risks should be assessed in that case.

13.3 An Alternative Approach

This short review has illustrated that the moral analysis of a minimal risk threshold in research is complex. However, since non-beneficial research is essential for improvements in the health of children as a group, determination of the morally appropriate threshold of risk for the conduct of non-beneficial paediatric research remains a practical need.

As we have seen, parents are morally justified in exposing their children to a certain level of risk; however, we have to consider the ethical justification of this exposure in each case before we reach conclusions concerning their exposure to research risks. The fact that risks exist in "daily life" or during "routine examinations" does not imply that the same level of risk should be morally acceptable in the research context; the concept of minimal risk is thus not only statistical, but also normative (Freedman et al. 1993; Nelson 2007).

Parents are also allowed to act beyond risks taken for their children's or the family's benefit, but also for the general public interest (e.g., charitable activities). A situation in which this is obvious is when parents decide to vaccinate their children against a disease, which does not constitute a significant threat to their children, so as to avoid its spread in the society; vaccination of males against rubella and of females against mumps are a couple of typical examples. Vaccination in this case exposes children to a (small) level of risk for unidentified others while offering no substantial personal benefit to all the individuals who are included in the vaccination programme.[6] These disease prevention strategies can only be successful if the individuals involved do not have purely self-interested motives (the whole population should be targeted to achieve population immunity).[7]

Moreover, since vaccination is a medical action it also shares some of the characteristics that daily activities or charitable activities do not necessarily possess, that is, unpleasantness, inconvenience, pain, fear to the child, and others (of course, in research settings these burdens or harms may be higher because some procedures must be repeated). As is the case with the "routine examinations standard", the proposed approach provides a direct analogy to benchmarking the probability and magnitude of harm or discomfort to which a child of a specific age will be exposed in a trial. The level of acceptable risks should therefore vary, depending on the particular age of the child (for instance the fact that an adolescent would feel less inconvenience than an infant from an injection should be taken into account in defining the notion of "minimal risk").

Another advantage of the proposed standard is that it avoids the problems of choosing between an objective and a relativistic interpretation. On the one hand, it

---

[6] Most people accept vaccines in situations in which the incidence of a vaccine-preventable disease is high, the disease is potentially serious and the risks from the vaccine are proportionately low (NCB 2007).

[7] Statistically, where there is fairly high vaccine coverage, the risks of disease for those who are unvaccinated may decrease (owing to population immunity) while the risks of vaccination remain (NCB 2007).

allows flexibility and adaptability, since it permits IRBs to take into account local conditions (i.e., seriousness and spread of the disease) and local moral standards (i.e., socially permissible risk in which parents are justified to expose their children for the social benefit).[8] On the other hand, the proposed standard could apply to both healthy and ill children without leading to exploitation or unfairness. The reasoning behind a parent's decision to vaccinate their ill or vulnerable child would not be that it is already exposed to a higher level of risk (due to its condition or the environment in which it lives). Parents (or legal guardians) are morally justified in refusing to vaccinate their vulnerable/ill children if the vaccine is considered dangerous for their children's condition. A parent's decision to vaccinate his or her child against a disease that does not pose a significant threat to the child points toward the maintenance of population immunity. Similarly, a parent's decision to enrol his or her child in a specific non-beneficial trial may be based on the probability of that child (or the child's future families) benefiting from the advanced medical care in the future.

Another analogy between parents' decision to enrol their children in a non-beneficial clinical study and their decision to vaccinate their children against an infectious disease is that both are intentionally chosen risk-laden activities and not unconscious or involuntary as are many decisions in daily life.

Finally, in both cases a "minor increase" over minimal risk could be morally justified when the potential benefit (immunisation against a serious disease in the case of vaccination and provision of essential knowledge for the subject's condition in the case of research) is of "vital importance".[9]

However, not all vaccines involve the same risks. Therefore, more empirical work is needed on the exact vaccination programmes that are appropriate to be used as a moral framework to assess minimal risk in non-beneficial paediatric research and for particular age groups.[10]

13.4 Conclusion

The purpose of this paper is to propose an alternative account by which the ethical threshold of acceptable risk in non-beneficial paediatric research can be assessed. We argue that consideration of the risk to which parents are willing to expose their children in a vaccination programme, as in the case of an epidemic, can facilitate the

---

[8] This is morally relevant since it could facilitate IRBs' ability to make qualitative assessments on the socially acceptable levels of intentionally imposed risk that each country would allow parents to apply to their children for the social benefit.

[9] This of course introduces a new risk threshold that must be determined, i.e., how much risk is justified by "vital importance"; however, the aim of this argument is to illustrate that the same moral justification for permitting an increase of the minimal risk threshold can be used in both cases.

[10] Although the proposed standard of minimal risk has the same limitations as the "daily life" standard (more empirical data is needed for its implementation), the empirical database in the proposed standard is still better and this makes its implementation easier.

definition of this threshold. Although the proposed account shares some of its strong points with already existing accounts, it does not lead to inconsistencies when applied in a research context, and as such it has the potential to contribute to the debate on the definition of minimal risk and facilitate the development and implementation of relevant policies.

References

Advisory Committee on Human Radiation Experiments (ACHRE). 1995. *Final report*. Bethesda: U.S. Government Printing Office.
Diekema, D.S., and F.B. Stapleton. 2006. Current controversies in pediatric research ethics: Proceedings introduction. *The Journal of Pediatrics* 149(1 Suppl): S1–S2.
Edwards, S., and J. Wilson. 2010. Hard paternalism, fairness and clinical research: Why not? *Bioethics* 26(2): 68–75.
Fisher, C.B., S.Z. Kornetsky, and E.D. Prentice. 2007. Determining risk in pediatric research with no prospect of direct benefit: Time for a national consensus on the interpretation of federal regulations. *American Journal of Bioethics* 7(3): 5–10.
Freedman, B., A. Fuks, and C. Weijer. 1993. In loco parentis minimal risk as ethical threshold for research upon children. *Hastings Center Report* 23(2): 13–19.
Glass, K.C., and M. Speyer-Ofenberg. 1996. Incompetent persons as research subjects and the ethics of minimal risk. *Cambridge Quarterly of Healthcare Ethics* 5(3): 362–372.
Glass, K.C., and A. Binik. 2008. Rethinking risk in pediatric research. *The Journal of Law, Medicine & Ethics* 36(3): 567–576.
Janofsky, J., and B. Starfield. 1981. Assessment of risk in research on children. *The Journal of Pediatrics* 98(5): 842–846.
Kopelman, L.M. 2004. Minimal risk as an international ethical standard in research. *Journal of Medicine and Philosophy* 29(3): 351–378.
Korenman, S.G. 2004. Research in children: Assessing risks and benefits. *Pediatric Research* 56: 165–166.
Litton, P. 2008. Non-beneficial pediatric research and the best interests standard: A legal and ethical reconciliation. *Yale Journal of Health Policy, Law, and Ethics* 8(2): 359–420.
Marshall, J.K. 2000. A critical approach to clinical practice guidelines. *Canadian Journal of Gastroenterology* 14(6): 505–509.
National Bioethics Advisory Commission (NBAC). 2001. *Ethical and policy issues in research involving human participants*. Report and recommendations of the National Bioethics Advisory Commission. Bethesda: U.S. Government Printing Office.
Nelson, M.R. 2007. Minimal risk, yet again. *The Journal of Pediatrics* 150(6): 570–572.
Nuffield Council on Bioethics (NCB). 2007. *Public health: Ethical issues*. Cambridge: Cambridge Publishers Ltd.
Resnik, D.B. 2005. Eliminating the daily life risks standard from the definition of minimal risk. *Journal of Medical Ethics* 31(1): 35–38.
Reynolds, W.W., and M.R. Nelson. 2008. Empirical data and the acceptability of research risk: A commentary on the charitable participation standard. *Archives of Pediatrics and Adolescent Medicine* 162(1): 88–90.
Rid, A., and D. Wendler. 2010. Risk-benefit assessment in medical research: Critical review and open questions. *Law, Probability and Risk* 9(3–4): 151–177.
Sharav, V.H. 2004. Alliance for human research protection: Author responds to letters on "children in clinical research: A conflict of moral values.". *American Journal of Bioethics* 4(3): 36–37.

Tan, J.O., and M. Koelch. 2008. The ethics of pharmacological research in legal minors. *Child & Adolescent Psychiatry & Mental Health* 2: 39.

Tracey, E.C. 2006. The search for minimal risk in international pediatric clinical trials. *Santa Clara Journal of Internationl Law* 5(1): 8–33.

Weijer, C. 2000. The ethical analysis of risk. *The Journal of Law, Medicine & Ethics* 28(4): 344–361.

Wendler, D. 2005. Protecting subjects who cannot give consent: Toward a better standard for "minimal" risks. *Hastings Center Report* 35(5): 37–43.

Wendler, D., and L. Glantz. 2007. A standard for assessing the risks of pediatric research: Pro and con. *The Journal of Pediatrics* 150(6): 579–582.

Chapter 14
Big Biobanks: Three Major Governance Challenges and Some Mini-constitutional Responses

Roger Brownsword

Abstract The development of "Big Biobanks" (population-wide biobanks that are established as a resource to be curated for access and use by the research community) is relatively new, and it is taking place at a time when the possibility of undertaking quite detailed genotyping and sequencing is assuming much greater prominence. Although there is much to debate concerning such biobanks, there is broad agreement that their good governance and legitimacy hinges on two fundamental conditions: first, that the interests of the participants are respected; and, secondly, that the activities are compatible with the public interest. Given this context, three of the many governance challenges faced by Big Biobanks will be discussed. First, there is question of whether individual "informed consent" can continue to function where hundreds of thousands of participants are involved and where the particular research purposes and projects to be pursued are not specified in advance. Secondly, there is the hot topic of the moment, namely whether biobanks have any responsibility to return individual clinically-significant findings to participants who, because of the longitudinal nature of such research, remain identifiable. Thirdly, there is the question of how the public interest is to be understood and applied: in which circumstances will access be denied as contrary to the public interest (even if the application is otherwise consistent with the consent given by participants) and, conversely, in which circumstances will access be granted for reasons of the public interest notwithstanding that the application is inconsistent with the consent given by participants?

This paper is a revised version of a talk that was first given at a workshop on "Consenting to Biobank Research" held at the University of Hannover in August 2013, and then of a lecture given at the University of Hong Kong in February 2014. I am grateful for comments made by participants at those events; but, of course, the usual disclaimers apply. I should make it absolutely clear that this paper is written in my personal capacity and not as Chair of the Ethics and Governance Council of UK Biobank (2011–2015).

R. Brownsword (✉)
The Dickson Poon School of Law, King's College London, Strand, London WC2R 2LS, UK
e-mail: roger.brownsword@kcl.ac.uk

14.1 Introduction

Broadly speaking, "biobanks" are to be understood as collections of biological samples and tissues that are curated and used for health-related research purposes. These common characteristics notwithstanding, biobanks display a considerable variety in their particular features (EC 2012a). For example, in some biobanks, the biological materials are complemented by various kinds of personal data (such as data concerning lifestyles) as well as medical records, while in others they are not; in some instances, the participants already have a diagnosis and are receiving treatment, while in others they are simply healthy volunteers who are recruited for the project; in some instances, the research purposes are quite specific, while in others the purposes (although health-related) are less well specified; in some biobanks, the resource is available only to a particular team of researchers, while in others it is open to the community of health researchers worldwide; in some cases, "for profit" applicants will be denied access, while in others they will not, and so on. In all cases, however, the prevailing view is that the good governance and legitimacy of the biobank and its research activities hinge on two fundamental conditions: first, that the interests of the participants (who are the sources of the samples, tissues and data) are respected; and, secondly, that the activities are compatible with, or not contrary to, the public interest (Brownsword 2013c).

In this paper, my focus is on "Big Biobanks"—that is, population-wide biobanks that are established (much like a library) as a resource to be curated for access and use by the research community. The development of such resources is relatively new and it is taking place at a time when the possibility of undertaking (and understanding the import of) quite detailed genotyping and sequencing is assuming much greater prominence (Brownsword 2012a). For example, in a recent report, the UK Human Genomics Strategy Group claims (HGSG 2012, 14):

> We are currently on the cusp of a revolution in healthcare: genomic medicine—patient diagnosis and treatment based on information about a person's entire DNA sequence, or "genome"—becoming part of mainstream healthcare practice. Increased knowledge and better use of genomic technologies and genetic data will form the basis for a reclassification of disease, with important implications both for predicting natural history and for identifying more effective therapies.

With the cost of sequencing falling rapidly—according to the HGSG, "it is not unrealistic to suggest that in a few years' time, we will be able to sequence a person's entire genome for the same cost, or less, than it currently costs to sequence a single gene" (HGSG 2012, 16)—and with the prospect of a beneficial transformation in healthcare, there surely will be significant investment in genomics research and, concomitantly, Big Biobanks.

Given this context, I will highlight three of the many challenges faced by the new generation of Big Biobanks—and, let me emphasise, there are many challenges: the three that I focus upon are by no means exhaustive. First, there is the much-debated question of whether individual "informed consent" can continue to function where hundreds of thousands of participants are involved and where the particular research purposes and projects to be pursued with the assistance of the biobank are not specified in advance. Secondly, there is the hot topic of the moment, namely whether biobanks (and researchers who have access to the resource) have any responsibility to return individual clinically significant findings to participants who, because of the longitudinal nature of such research, remain identifiable. Thirdly, so long as Big Biobanks aspire to function in the "public interest", there is the question of how that much-debated notion is understood and applied (Brownsword 1993). One question is: in which circumstances will access be denied as contrary to the public interest, notwithstanding that the application is otherwise consistent with the consent (authorisation) given by participants? And the partner question is: in which circumstances will access be granted for reasons of the public interest notwithstanding that the application is inconsistent with the consent (authorisation) given by participants?

14.2 Informed Consent, Broad Consent, and Variation of the Rules

Elsewhere, I have argued that there is much work to be done in improving our understanding of informed consent in our modern information societies (Brownsword 2012b). The key challenges include: clarifying the set of informational rights that we recognise—we need to have a clear view of privacy, confidentiality, data protection, the right to know, the right not to know, and so on; being more disciplined in relating informed consent to an ethic of rights—we should constantly remind ourselves that it is through the process of informed consent that we authorise acts that would otherwise infringe our rights (and thus permit a change of position relative to these rights); being more focused in differentiating between the information that A is entitled to have qua rights-holder and the sense in which A must be informed before A can give a valid consent relative to A's rights; and fully appreciating the important place of consent in binding parties to an agreed set of rules (whether these are the rules of a game or competition, the rules of a club or the "house" rules, or the rules of a business association, and so on). If we take these challenges on board, we will see that, pace its critics, informed consent should remain a central regulative principle, not only in the information society, but also in a world of Big Biobanks.

Bearing in mind these points, we can begin to get some clarity about the supposed need for Big Biobanks to abandon informed consent in favour of broad or generic consent. This proposition, I will suggest, is apt to mislead, by eliding the conditions for a valid consent with the breadth of the authorisation given. Then we can open up the question of how we should characterise the relationship between Big Biobanks and their participants. Traditionally, we understand such a relationship in tort-like terms (consent is a defence to what would otherwise be a tort, or even a crime); and, more recently, it has been suggested that we might view it instead as "contractual". However, I will suggest that the better view is that the terms and conditions for participation in Big Biobanks should be treated as akin to mini-constitutions. The constitutional provisions bind participants (and researchers) in the way that club rules bind members—on the basis of an originating informed consent (Beyleveld and Brownsword 2007). And, importantly, where revisions to the terms and conditions are made in whatever way is provided for by the constitution, participants (so long as they remain as participants) are bound by the revised terms and conditions because their originating consent has authorised this scheme of governance.

14.2.1 *Informed Consent and Broad Consent*

Is it correct to assert, as it is commonly asserted, that biobanking projects cannot operate with informed consent but, instead, need to have broad or generic (but not informed) consent from participants?[1] The short answer is that it is not: this simply confuses the authorisation of the act with the particular scope of the authorisation. Once this distinction has been clarified, it is obvious that what biobankers require is that their participants give a broad authorisation for their data and samples to be used in the pursuit of health-related research projects. If the researchers cannot say at the point of enrolment that the project will pursue only such and such specified research purposes, they need participants to grant broad consents that license the use of samples and data for a wide range of purposes—biobankers, quite understandably, do not want to have to keep coming back to participants to vary the consent. However, it is a mistake to think that this has a negative bearing on the need for informed consent at the point of enrolment.[2]

Unless we think that agents have a responsibility to assist biobanking projects, the starting position is that agents have a right not to participate, a right not to give samples, and various informational rights in relation to their health and lifestyle. For biobankers to act in ways that would otherwise violate these rights, they need the informed consent of their participants. Once A has agreed to participate, there is then a further question about the scope of the authorisation to be given to the researchers—which is where so-called broad consent enters the picture. If the biobank specifies that one of the non-negotiable terms and conditions for participation is that broad authorisation must be given, then participants must decide whether they wish to proceed. If they proceed, they do so on the basis of their originating informed consent; if they do not wish to participate on such terms, they walk away. In short, the consent is informed, the authorisation broad.

---

[1] For discussion of the consent options, and then advocacy of a model of "intermediate consent", see Forgó et al. 2010.

[2] In order to avoid any doubt about the legality of broad consent, Finland recently enacted a Biobank Act (Act 688/2012) that makes it clear that a participant may consent to a broad range of purposes; see Soini 2013. According to Joanna Stjernschantz Forsberg and Sirpa Soini, the authorisation may "include research into health-promoting activities, causes of disease, and disease prevention and treatment, as well as research and development projects that serve healthcare"; see Forsberg and Soini 2014. However, compare the restrictive Lifegene decision in Sweden (FN 3).

Or again, consider the risk of the samples and data that one provides being misused or applied in ways that have negative consequences for the particular participant. Recently, the Personal Genome Project-UK (PGP-UK), the fourth of its kind in the world, has launched on terms that are likely to be extremely unattractive to some potential participants (Sample 2013). Indeed, it will take a special kind of person to sign up for a project that explicitly says that, while participants "are not likely to benefit in any way as a result of [their] participation" (PGP-UK 2014, Art. 7.1), there are numerous potential risks and discomforts arising from participation (PGP-UK 2014, Art. 6). The consent to be given by participants is "open", in the sense that all medical information attached to a person's record will be made available online; and, while participants' names and addresses will not be advertised, participants are warned explicitly that they might quite easily be identified and their privacy cannot be guaranteed (PGP-UK 2014, Art. 6.1.a.iv). Moreover, prospective participants are put on notice that PGP-UK "cannot predict all of the risks, or the severity of the risks, that the public availability of [participant] information may pose to [participants] and [their] relatives." (PGP-UK 2014, Art. 6.1.a.vi) Clearly, this is not for everyone, and especially not for the risk-averse. However, if a participant signs up for these terms on a free and informed basis, then their open consent, just like a broad consent, authorises certain research activities; and they have accepted the risks that are within the scope of the authorisation.

That said, it might be objected that this analysis misses the point. The point, it might be said, is that prospective participants, faced with a non-negotiable term for broad authorisation, might decide not to participate; and that, if this happens in too many cases, biobanks will not be able to recruit sufficient numbers of participants. In a community of rights, unless we think that there is a responsibility to participate, this indeed is what it might mean if rights are taken seriously. However, the broadside on this outcome should be focused not so much on informed consent as on the covering rights.[3] If that really is the objection, then we need to see it for what it is—a utilitarian attack on rights, not a rights-based reservation about informed consent (Brownsword 2009).

---

[3] There is also a potential difficulty for researchers if the informational rights recognised in relation to the processing of personal data build in a "specific purposes" limitation. In principle, a rights-holder might give an informed consent allowing researchers to process the data for broad purposes; but, in practice, there would need to be very explicit signalling of such an authorisation. So, for example, in a case involving the biobank Lifegene, the Director of Sweden's Data Inspection Agency has ruled that the gathering of personal information for "future research" is in breach of the Personal Data Act. The problem is that Lifegene's expressed research purposes are too general to satisfy section 9c of the Act, which provides that the collection and processing of personal information must be for "specific, explicitly stated and justified purposes […]." I am indebted to Adrienne Hunt for drawing my attention to this case and to Søren Holm for translating and summarising the case. See: http://ethicsblog.crb.uu.se/2011/12/20/the-swedish-data-inspection-board-stops-large-biobank/. Accessed 14 Feb 2015.

Of course, this is now all subject to the fate of the European Commission's proposed General Data Protection Regulation: see EC 2012b.

14.2.2 Varying the Rules

One of the attractions of broad or generic consent (or, more accurately, informed consent with a broad authorisation) is that there will be fewer (if any) occasions when it is necessary for a Big Biobank to return to its participants to get them re-consented for projects or purposes that are outwith the original consents. However, in practice, all authorisations have their limits and, even with broad consents in place, the need for a Big Biobank to seek fresh consents for an otherwise unauthorised research use cannot be ruled out. This prompts the thought that it would facilitate the operation of Big Biobanks if they could more readily vary their terms and conditions without this being contrary to the consents given by participants. In order to be in a position to address this possibility, we need to think about the best way of characterising the relationship between a Big Biobank and its participants.

Traditionally, the relationship between researchers and participants is largely modelled on that between clinicians and patients—and, perhaps understandably so, given that, in practice, the line between research and clinical treatment is not entirely clear-cut.[4] At all events, the model in question is one of individual "informed consent" (WMA 2013, 18.25–32). It is for each individual patient to authorise the clinician to undertake a particular treatment regime; it is for each participant to authorise the acts of the researchers; and, in the absence of such authorisation, there is a prima facie wrong. Let me call this the "tort model". The first characteristic of this model is that it is for each individual right-holder to give the controlling authorisation. There is no question of decisions being made by groups, where majoritarian principles prevail. Each individual has a veto; if the individual says "no" to the researcher or to the clinician or to the proposed treatment, then that bars the researcher or the clinician from proceeding. Secondly, the model is one of authorisation: by giving consent, the participant or patient waives the benefit of some protected right or interest—for example, by consenting to the taking of blood, the participant or patient authorises an act that would otherwise constitute an assault. Thirdly, the consent will be valid only if it is given on a free and informed basis. Specifying the conditions for a "free" and an "informed" consent is a minefield of difficulties. However, paradigmatically, a consent is not free if it is obtained by coercion, and it is not informed if it is procured by fraud (Beyleveld and Brownsword 2007).

Although critics argue that the problem with this traditional tort model is that its informed consent requirement does not copy across to Big Biobanks where the many different future research uses cannot be specified, this (as I have already explained) trades on some confusion. Participants in such biobanking projects can give perfectly valid informed consents even though the projected research uses are largely unspecified. However, for a quite different reason, the tort model is

---

[4] Compare the way in which clinical care slides into research in the early clauses of the current version of the *WMA Declaration of Helsinki. Ethical Principles for Medical Research Involving Human Subjects* (WMA 2013).

problematic. The reason is that each individual participant has a veto. At the stage of enrolment, it is entirely appropriate that each prospective participant should be able to say "no"—and this might mean that they are precluded from participating. It is also appropriate that each participant should have the right to withdraw. However, where a Big Biobank wants to modify the terms of participation, it is arguable that its governance needs a more flexible model, one that does not require each and every participant to authorise the proposed variation.

A second model that seems to work better with Big Biobanks is one that is "contractual". Even if the agreement between the biobank and its participants would not be treated as a legally enforceable contract (because it is not backed by the requisite intention),[5] the relationship is nevertheless contractual. What this means is that the participants sign up to a package of terms and conditions. So, for example, when blood is taken from a participant, the authorisation is not (as per the tort model) in the individual interaction between researcher and participant but in the agreement to participate on terms and conditions that contemplate blood being taken. This in no sense abandons informed consent as the justifying basis, because the consent to the background terms and conditions (that give the many particular authorisations to the researchers) needs to be free and informed. If participants are coerced into signing up to the package, or if they sign up on the basis of a fraudulent prospectus, their consent is invalid and the researchers will be disallowed from relying on the supposed authorisation.

The contractual model fits quite neatly with enrolment in a Big Biobank. Participants come to the table with many protected rights or interests (in their person, in their property, in their privacy, and so on) and, rather than giving ad hoc consent as each right is engaged, they consent to the authorising package. Potentially, the contractual model also allows for some flexibility to be built into the future governance of the biobank. In many standard form contracts (both commercial and consumer) there will be clauses that provide for some variation in the performance. If clauses of this kind were included in the package to which participants are invited to agree, this might authorise Big Biobanks to operate with more flexible governance arrangements. Of course, prospective participants might be put off by clauses of this kind; and, actual participants might so disapprove of some variation that they decide to withdraw. However, if participants are willing to subject themselves to such an arrangement, all is well; and this is contractual business as usual.

There is a third way in which we might model the relationship between Big Biobanks and their participants. The fact that we might be a bit uncomfortable with a contractual model—which might sound rather too close to the market and not

---

[5] In the English law of contract, an independent requirement of an "intention to create legal [contractual] relations" was introduced by the Court of Appeal in *Balfour v Balfour* [1919] 2 K.B. 571. This requirement translates into a rebuttable presumption that domestic and social agreements are not backed by the requisite intention and thus should not be treated as legally enforceable contracts; by contrast, the presumption is that business agreements are backed by an intention to create legal relations and are enforceable. It is not clear how an otherwise contractual relationship between researchers and participants would be classified; but, my guess is that the presumption would be that the parties do not intend to create a legally enforceable agreement.

sufficiently aligned with the altruistic culture of much participation (NCB 2011)—points us instead in the direction of clubs and associations. Suppose that we view the relationship between participants and Big Biobanks as akin to membership of a club; the participants are club members; the "package" is not so much a contract as a (mini) constitution. In other words, the governance frameworks (the packages) that regulate the relationship between Big Biobanks and their participants are to be understood as constitutional documents. Again, there is no weakening of the underlying justificatory role of informed consent: participants only become members of these clubs, subject to their constitutional arrangements, if they do so on a free and informed basis. Crucially, the constitution may provide for variation of the terms and conditions in many different ways, including by notice and comment procedures, by majority-voting, by consultation and deliberative democratic decision-making, and so on.[6] Whatever the process, it does not have to give each member a veto as in the traditional tort model or, indeed, require a response from each participant.[7]

Finally, it is worth observing that the constitutional model—which is surely the way that we should now understand the relationship between Big Biobanks and their participants—has a capacity for ongoing adjustment that might attract proponents of various "dynamic" models of consent (EC 2012a; Kaye et al. 2011). The idea of these models is that modern on-line technologies should be utilised so that participants, having initially given rather narrow, specific consents, are then invited actively to opt-in for secondary or downstream research uses. Critics point out that dynamic consent, so conceived, is not an unqualified good. For example, Steinsbekk, Kåre Myskja, and Solberg conclude (Steinsbekk et al. 2013, 901):

> The dynamic consent strategy with repeatedly opt-in options holds the risk of participants not opting in or opting in with a bad conscience for not making an informed choice, risk of weaker ethical review of research projects, risk of disillusionment based on unfulfilled expectations, as well as the risk of inviting participants into therapeutic misconception.

To some extent, these perceived risks arise because the dynamic approach is treated as a gloss on the traditional tort model. However, once we start thinking in terms of the constitutional model, we reduce these risks. No longer is the evolving governance of Big Biobanks constrained by "individualised" decision-making; instead,

---

[6] Article IVc of the Ethics and Governance Framework of UK Biobank (2007) provides for revision in the following somewhat general terms:

> The Board of Directors, the Ethics and Governance Council, the Funders and other interested parties (including participants and members of the wider public) may propose amendments or revisions of the Framework. In particular, the Ethics and Governance Council will advise on outstanding issues, and may propose adjustments in response to new developments. Adoption of any amendment or revision will rest with the Board of Directors.

[7] Compare Article 3.4 of the PGP-UK consent form (PGP-UK 2014). This provides that proposed revisions to the terms and conditions have to be first reviewed and approved by the relevant research ethics committee. Then, each participant will be asked to sign up to the revisions. Those who sign up will continue as participants; those who do not will have their accounts deactivated until such time as they review and sign up to the revisions.

governance is based on a quite different, associative, way of understanding the relationship between Big Biobanks and their participants.

14.2.3 Taking Stock

In an interesting paper concerning the particular respects in which it is more and less important for participants to have information about the specific research purposes that are proposed, Matteo Macilotti describes the context as one in which "from consent on the specific research project, we are moving towards consent on a model of governance [meaning broad and blanket consent]." (Macilotti 2013, 144) This very nearly hits the nail on the head. We are moving from a model of informed consent that authorises use of data and samples for a specific, named research project to a model that authorises use for a broad range of (as yet) unascertained research purposes. To some extent, the latter does represent "a model of governance" but it still presupposes a tort-type relationship between the biobank and participants. The model of governance that Big Biobanks require is one that presupposes an associational model with a flexible governance framework viewed as a mini-constitution. Within this constitutional package, there might be some authorisations that are quite specific alongside others that are much more general; but there might also be agreed authorising procedures (not necessarily requiring a "yes" vote from each participant) to be employed when the biobank needs a fresh mandate for its activities. It is when participants, through their originating consent, sign up to arrangements of this kind that we have the desired move to "consent to a model of governance".

14.3 The Responsibility to Give Feedback

Currently, one of the most complex and contested issues in the ethics and governance of biobanks—a question that Catherine Heeney and Michael Parker rightly single out as "[o]ne of the most hotly debated" in the context of modern biobanking practice—is "whether there is an obligation to feedback research results to participants." (Heeney and Parker 2012) Suppose, for example, that researchers conducting genetic analysis on biobanked materials identify a particular mutation for breast cancer in a sample provided by an identifiable participant. Do the researchers have an obligation to inform the participant; or, to turn this round, does the participant have a right to be informed? If it is claimed that the participant does have such a right, a host of further questions need to be addressed, including questions about the scope of the right, its weight in relation to any competing rights, whether researchers owe feedback responsibilities also to third parties (such as relatives of the participant), how the information is to be conveyed to the participant, how the right might be affected by an explicit "no feedback" policy at the biobank, and whether the right to be informed also implies a right that researchers actively "look out for" potentially clinically significant findings (Gliwa and Berkman 2013).

In what follows, I will not attempt to stake out a position in what is an extremely difficult debate.[8] In that debate there are disagreements about whether feedback should be governed by a calculation of net benefit to participants or by a "right to know" or by some plurality of principles[9]; and, even if the governing principle is agreed, there might be quite significant disagreements about its application in particular cases—for example, there might be different views about whether a particular rule for feedback will generate more net benefit than a rival rule; or of course, there might be disagreements about the balance of benefit and harm in a particular case.[10] In this paper, my intentions are somewhat modest: I simply want to draw attention to an important ambiguity in the declaration of a "no feedback" policy; and then I will explore the basis on which a participant might claim to have, not merely an expectation, but a reasonable expectation that there will be feedback of individual findings that are potentially of clinical significance.

14.3.1 The Meaning of a "No Feedback" Rule

In many biobanks, the general rule is that "no feedback" will be given to participants. However, in two respects "no feedback" is open to misunderstanding. The "no feedback" notice notwithstanding, there might actually be some findings that are returned to participants.

First, even where the declared policy is one of "no feedback", this usually refers to the position once participants' samples and data have been "banked". Prior to that point, some health-related information might be given to participants. For example, at the point of enrolment, participants might be given their blood pressure or bone

---

[8] Compare the Presidential Commission for the Study of Bioethical Issues, Anticipate and Communicate—Ethical Management of Incidental and Secondary Findings in the Clinical, Research, and Direct-to-Consumer Contexts (PCSBI 2013, 3):

> The current challenge for public policy and professional ethics is to identify through thoughtful deliberation specific criteria that practitioners can use to determine when it is ethically permissible or obligatory for clinicians, researchers, or DTC companies to disclose and not to disclose incidental findings to patients, participants, or consumers.

[9] The Presidential Commission (PCSBI 2013, 4) identifies "four ethical principles to be particularly applicable to the ethical assessment of incidental and secondary findings: respect for persons, beneficence, justice and fairness, and intellectual freedom and responsibility." Compare the criteria proposed by Kaye et al. 2014. (But, nb, the authors' caution that their approach does not translate straightforwardly to large population-based biobanks.)

[10] In this light, we should note the Presidential Commission's remarks about the need for more empirical research concerning the impact of giving feedback and the attitudes of participants towards the return of findings (see, e.g., PCSBI 2013, 7, rec 3).

density readings, or their BMI score, and so on[11]; and there might also be some advice about incidental observations, as when a participant might be advised to check out a suspicious-looking mole with their doctor. However, once the samples and data have been collected and "banked" for the use of researchers, the "no feedback" policy signals that the general rule is that there will be no individual feedback arising from the findings made by researchers (general findings, of course, will be disseminated in the usual way).

The second misunderstanding is more subtle. When a biobank declares that its policy is "no feedback", this might mean quite literally that there is no feedback, that there is an absolute rule against feedback. Here, "no feedback" is intended to signify that participants have no right to feedback and that researchers have a duty not to inform; in no circumstances will even clinically significant results of research undertaken on the banked materials be returned to individual participants. However, "no feedback" might signal something more specific, namely that the researchers do not accept any obligation to give feedback. This latter reading is designed to counter participants who assert that they have a right to be informed and that, concomitantly, the researchers have matching obligations. While such a reading is intended to shield researchers against claims for feedback made by participants, unlike the first reading, it does not preclude the giving of feedback; but, whether or not participants are informed, is exclusively for the biobanks and their researchers to decide—for example, by returning findings only where the condition is serious and treatable, or where there is a clear balance of benefit to the participant.

We can allow that, in general, a "no feedback" policy is motivated by the best paternalistic intentions, by a concern about false alarms and causing unnecessary distress to participants. However, in an age when paternalism is no longer the governing approach for clinicians in their relationship with patients, is it a defensible approach for researchers at Big Biobanks? Heeney and Parker come close to answering this question. Noting that one "no feedback" strategy is, in effect, to manage the expectations of participants, they say (Heeney and Parker 2012, 296):

> One route would be to make it clear to participants and health professionals at the time of consent that there will be no feedback of research results. There are a number of arguments supporting this including its potential for greater clarity about consent and about the distinction between research and clinical care and the fact that feedback assumes some sort of infrastructure in which the connection with participants is maintained to the extent that they can still be contacted and told to seek medical advice, for example.

However, as the authors remark, this approach has fallen out of favour in those cases where research might "produce very clear evidence of a serious harm which might be avoided by an easily available intervention and where there exists something akin to a duty of easy rescue." (Heeney and Parker 2012) In other words, if our premise is that participants have a positive right to be informed (possibly akin to a positive right to be rescued where this is straightforward and not difficult for the rescuer), we are less likely to judge that biobanks do the right thing by withholding

---

[11] For the view that "raw personal data" should be accessible to individual participants, see Lunshof et al. 2014.

clinically significant findings. Moreover, even if a "no feedback" policy has been "communicated" to participants, we might wonder whether its significance has been fully appreciated; and we might judge that, regardless of the biobank's declared policy, participants "reasonably expect" to be given feedback where a biobank holds clinically significant, serious, and actionable information about a particular individual (Knoppers et al. 2013; Wolf et al. 2012).

These observations invite some further thoughts about the basis on which participants might claim to have a reasonable expectation of feedback (Brownsword 2007).

14.3.2 The "Reasonable Expectations" of Participants

One of the striking features of much of our ethical and regulatory thinking (notably in relation to the interest in privacy) is that the recognition of a right hinges on the question of whether we judge that a person has a "reasonable expectation" that his or her particular interests will be respected and protected (Brownsword 2012c). If a participant's claim to have feedback hinges on whether it is based on a reasonable expectation to have feedback, then the question is: by reference to what standard or practice or to whose authority is the expectation judged to be a reasonable one?

First, the participant might invoke relevant background rules of law. There has been much discussion of whether a participant might succeed against a researcher in a tort claim for wrongful non-disclosure (Johnston and Kaye 2004). The consensus is that English law does not clearly support such a claim; and, if the legal test turns on whether it is "fair, just, and reasonable" to place researchers under a feedback responsibility, this seems merely to restate the original question of whether the claimant's expectation is a reasonable one.

Where a biobank declares a "no feedback" policy which is clearly notified to participants, if anything, this further weakens the participant's tort claim. However, where "no feedback" signals that the biobank reserves a discretion to give feedback, a participant might argue in a judicial review that the discretion has been exercised improperly. If successful, the claimant might compel the researchers or biobank to reconsider their decision; but, of course, a claim of this kind would only get off the ground if the policy set by the biobank and its administration were recognised as having a sufficient "public" character to render it susceptible in principle to judicial review.[12]

Secondly, the participant might claim that the researchers had formally or informally signalled that feedback would be given. Where a biobank has sought to manage participants' expectations by having a well-advertised "no feedback" policy, this

---

[12] In English law, the rules of the biobank might be likened to those of a private club and, as such, not judicially reviewable (notwithstanding the "public" dimensions of a big biobank). Compare, e.g., R v Disciplinary Committee of the Jockey Club, ex parte Aga Khan [1993] 2 All ER 833.

kind of claim would be unlikely to succeed. However, if the policy had not been signalled, then the claim would turn on whether the participant could show on the facts that an undertaking to provide feedback had been given. In principle, unless the background law prohibited researchers from giving feedback, their voluntary assumption of a responsibility to give feedback would be a strong ground for claiming a reasonable expectation of feedback.

Thirdly, the participant might rely on a general attitude that there should be some reciprocity in the relationship with researchers: participants assist researchers in various ways in return for which researchers should assist participants by giving appropriate feedback. There does seem to be evidence that at least some (and, quite possibly, many) participants sign up with the expectation that there will be reciprocation (Beskow et al. 2011; Bovenberg et al. 2009; WTMRC 2012). However, the fact that others share one's own expectation does not make anyone's expectation reasonable. Possibly, the claim for reciprocity might be grounded in some other way—for example, in the way that Henry Richardson relies on the relationship of "entrustment" between participants and researchers (Richardson 2012); but it is not enough that the de facto expectation is widely held by participants.

Fourthly, the participant might rely on the settled custom and practice at other biobanks or in a certain sector of research (or, indeed, in clinical practice as genetic analysis becomes routine). For example, it might be that researchers who work with MRI scans might consider it best practice to return incidental findings to their participants. Accordingly, where there are such practices and where the claimant participant is dealing with researchers at a biobank with no declared policy on feedback, the unstated assumption (and expectation) that there will be feedback might look perfectly reasonable. However, where "no feedback" is the declared rule, contrary custom and practice notwithstanding, the argument that there is a reasonable expectation of feedback is seriously weakened.

What these appeals to reasonableness have in common is that they rely on a range of contingent factors being set in the right way. If the law supports a claim to feedback, if researchers voluntarily assume a responsibility to give feedback, if custom and practice supports giving feedback, and the like, then the participant's claim will get to first base; and, other things (such as the notification of the biobank's policy) being equal, the participant's expectation will show as a reasonable one. Where "reasonable expectation" is the test, then—in appropriate cases—the participant will be judged to be entitled to be informed.

There remains the possibility that a participant might claim to have a reasonable expectation of being given feedback because a right to be informed is grounded in reason—not in contingent legal provisions or promises or custom and practice. Quite simply, if a participant has such a reason-based right to be informed, it would be reasonable (to put the claim at its lowest) to expect to be informed. However, in an age of deep scepticism about such claims, the $64,000 question, which I will not attempt to respond to here, is whether, and if so how, such a right might be rationally grounded (Beyleveld and Brownsword 2015).

14.4 Public Interest

In this final part of the paper, I turn to the concept of the public interest. We start by considering the complaint that informed consent is an unnecessary transaction cost that impedes prospectively beneficial research and, thus, operates in a way that is contrary to the public interest. Then, we consider some scenarios in which an appeal is made to the public interest in order to justify granting access in circumstances where this would seem to be beyond, or even directly contrary to, the authorisations given by the participants in their consent. This raises some important questions about the coherence and consistency of a biobank's understanding and application of the concept of the public interest.

14.4.1 Health Research and Informed Consent

Let us suppose that health research, if not an unqualified good, is at least a prima facie good. Governments that invest in and encourage such research would not normally have to defend the propriety of their purposes; broadly speaking, public health aligns with the public interest in the sense that its promotion is suitably a public matter as well as generally beneficial. Nevertheless, the idea that health researchers should be permitted to by-pass informed consent is a very dangerous one. Where physical rights are at stake, I take it that everyone agrees that researchers should not be able to conscript subjects for trials, or commandeer organs for research, or even kill subjects in order to advance their understanding, without the subjects being willing to cooperate and without their giving their informed consent. If there were any doubt about this, the UNESCO Universal Declaration on Bioethics and Human Rights proclaims again that "the interests and welfare of the individual should have priority over the sole interest of science or society." (UNESCO 2005; likewise see: WMA 2013). Moreover, I take it that we would not accept an opt-out regime as being sufficient to license conscription and cooperation. Yet, where the relevant background interests are informational, where researchers want to access medical records or other health-related information, why should we think that these ground rules should be changed?

One reason for thinking that these ground rules should be changed is that we judge that, where (mere) informational interests are at stake, the balance of benefits and burdens swings strongly towards the interests of researchers. However, to argue in this vein that consent interferes with legitimate public interest purposes is to beg the question in favour of utilitarianism and to misrepresent the place and significance of informed consent. In Europe, with its high-profile commitment to respect for human rights (and underlying this, human dignity) (cf. Brownsword 2013b; Düwell et al. 2014), the context that we should presuppose is one of a community that takes individual informational rights (and rights more generally) seriously. The question is not whether informed consent is a gratuitous obstacle to research in a society that is guided by a utilitarian outlook but whether it is a problem for a community of rights that has developed modern information technologies (Brownsword 2008; Brownsword and Goodwin 2012).

In a community of rights, public interest considerations will help to define the shape and scope of individual rights (Beyleveld 2006); and, as we have said, there is much work to be done in sharpening up our thinking about the informational rights that we recognise both off-line and on-line. If, on analysis, we judge that no right is engaged, consent simply is not an issue—to reason otherwise is to commit the Fallacy of Necessity (Brownsword 2004). If, for example, we judge that information that concerns us, once anonymised, engages no informational rights, then researchers may use such information without getting covering consents. Indeed, as I have argued elsewhere, there might be cases in which we have positive responsibilities to assist researchers (Brownsword 2009). Even without that, there might also be cases (driven by more compelling rights than whatever informational rights are at stake) where researchers who press ahead without getting the informed consent of the relevant rights-holders might still be justified all things considered. For example, there might be a case where one of the informational rights is overridden by the conflicting right to life of an agent; and there might also be cases in which one informational right is overridden for the sake of another (more compelling) *informational* right—for example, where there is a conflict between confidentiality and the right to know.

None of this is suggesting that informed consent is a straightforward regulative principle. However, as a regulative principle, it must be mapped within a framework of rights; and it is simply inappropriate to try to dislodge it, and the covering rights, by appealing to the public interest (wherever there is a perceived public benefit) while silently presupposing a utilitarian regulatory environment.

14.4.2 Participants' Consent and the Public Interest

Here, my interest is in exploring one of the potential tensions between respecting the interests of participants (whatever this might mean) and acting in line with the public interest (however this concept is understood). In principle, the public interest might be relied on as a reason for denying access to a Big Biobank even though the application does not seem to violate the interests of participants—for example, it might be argued that even though a tobacco company's application is consistent with the participants' consent, in the sense that it is bona fide and for health-related reasons, it would be contrary to the public interest to grant access; or, the public interest might be invoked as a reason for granting access even though the application is not authorised by the participants' consent. It is this latter use of the public interest that I will focus on.

Let us assume a Big Biobank where the resource is open to researchers beyond those who have the primary responsibility for curating the collection. When such researchers apply for access to the biobank, an access committee adjudicates the application. According to the governance framework (the "constitution" as I would have it), two of the questions that the committee must ask are: first, whether the application proposes a use of the resource that would be incompatible with the interests of participants; and, secondly, whether the application should be refused on the ground that it is in some respect contrary to the public interest. In the ideal-typical case, access to biobanks will be granted to researcher-applicants, whose health-related projects are clearly in the public interest and where granting access is plainly compatible with respecting the interests of participants. Conversely, access to biobanks should be denied where the applicants' purposes are not in the public interest and where granting access would be incompatible with the interests of participants. However, what about those applications that are less clear cut? For example, what if the application, although it is judged to be in the public interest, is incompatible with the interests of participants? In these more difficult cases, is there any clear priority as between the interests of the participants and the public interest; and, if there is, what is it?

Consider, first, a scenario where there is an application to a Big Biobank, Biobank #1, to link data in Biobank #1 with data in Biobank #2, in each case the data relating to the same group of individuals (these individuals being participants in both biobanks). The linkage promises to improve the power of the research. However, it is not authorised by the consents taken from the participants. Each Biobank has its own set of terms and conditions and there is no provision for linking data or sharing samples with other biobanks. Nevertheless, it is tempting to argue that this linkage is in the public interest and so it overrides any constraints set by the participants' consents. There might also be the gloss that, if the participants were asked to give their consent to this proposed linkage, they would give the necessary authorisations to each Biobank. However, this gloss invites the retort that if, were they to be asked, the participants would consent then why not ask them and get the necessary individual consents? To be sure, this involves some inconvenience but to appeal to the public interest (qua public benefit) to justify avoiding a transaction or opportunity cost looks like a reversion to utilitarianism—which, as I have said, simply will not do in a community of rights.

Secondly, consider a test-case of the kind that occurred several years ago in Sweden, where an application was made to a biobank for the purpose of assisting with the identification of Swedes who were victims of the Boxing Day Tsunami. If the governance framework contemplates access only for "health-related research purposes", even an imaginative lawyer would have difficulty in construing this as covering the identification of victims of disasters. On the other hand, the public interest argument is attractive and it is plausible to suppose (as, indeed, was the case in Sweden) that there would be broad support for access in such circumstances.

At much the same time, a third test-case also arose in Sweden when the police were given access to a biobank to assist with their inquiries into the murder of the politician Anna Lindh. Here, although it is hard to gainsay the public interest in the prevention and detection of crime, and although this was the most serious of crimes, the popular view was that access should not have been granted. If our hypothetical Big Biobank, in company with many biobanks, has given categorical assurances to participants that access will not be granted to the police (or to insurance companies or employers), then this is a clear breach of the biobank constitution. However, we might wonder whether it can be right that promises made to participants have greater weight than the community's interest in the conviction and punishment of murderers.

Clearly, if we are to develop a principled interpretation of our practice or if we are to improve on our intuitionistic responses to difficult cases where the private interests of participants and the public interest are in tension, there is some major theory-building to be undertaken (Capps 2013). In the best of worlds, we would have a clear understanding of the constituent elements of the public interest, particularly of those elements that confine and constrain actions even though they might be "beneficial" in a general sense; we would have a clear appreciation of which matters of governance are for public determination and which for private decision; and we would have a defensible overarching framework that would relate the private interests of, and the consents given by, participants to the more general public interest. Within the space for legitimate private governance that is accorded by this framework, Big Biobanks would be permitted to make express provision in their mini-constitutions for dealing with unforeseen access applications that are either within the spirit (if not the letter) of the project or that serve some other aspect of the public interest.

In the current under-developed state of our understanding, let me suggest that in a community of rights "the public interest" (whether as a reason for denying or for granting access) would be understood as taking its place within a hierarchy of reasons.[13] The hierarchy would comprise four classes of reason: namely, that access should be granted or denied in order

- to protect the essential infrastructural conditions on which the existence of the community is predicated
- to protect and respect fundamental rights (in particular, cosmopolitan guarantees of respect for human rights and human dignity)
- to protect or serve the public interest
- to respect the Biobank's self-governing scheme (its mini-constitution).

Each of these classes of reason (infrastructural catastrophe, fundamental rights, public interest, and private codes) invites further analysis and elaboration. However, for present purposes, the important point to emphasise is their relative exclusionary effects. These effects are as follows: (i) where considerations of the public interest are in tension with the application of the Biobank's mini-constitution, the former will prevail (i.e. exclude the latter); (ii) where different strands of the public interest are in competition with one another, they do not exclude one another but a judgment will have to be made about where the balance of interest lies—the conclusion being expressed in terms of the public interest requiring such and such actions to be taken; (iii) where the pursuit of the public interest is incompatible with respect for

---

[13] For my conception of a "community of rights", see e.g. Brownsword 2008; and, for a related analysis about the ordering of a community, see Brownsword 2013a.

fundamental rights, the latter prevail (i.e. exclude the former); and (iv) where there is a danger of catastrophic infrastructural failure, this provides a reason for precautionary action that overrides all other reasons (i.e. it is comprehensively exclusionary). With this scheme in the background, we might offer the following thoughts on the three test-cases that we have sketched.

First, there is the proposed linkage of data between different biobanks. Now, while it might be said to be in the "common interest" of the participants that the linkage should take place, this is not quite the same as saying that it is in the public interest (Bell 1993; Milne 1993). However, it might also be argued to be in the public interest in the sense that linkage of this kind increases the power of the data and, thus, promises to further promote health-care research. If this argument is accepted, there is a public interest reason to exclude (prevail over) whatever restrictions on access have been imposed by the Biobank's private governance scheme (its mini-constitution). However, this is not quite the end of the matter. For, even if there are not considerations of fundamental rights or catastrophe that trump the public interest argument, there might be other public interest considerations in play. For example, there might be a concern that, in the longer run, the public interest in research will be damaged if the terms of biobank governance are set aside so readily—because participants might be reluctant to come forward if they fear that the mini-constitution is liable to be set aside so easily.

That said, the particular mini-constitution might provide considerable flexibility by providing a special process where access is for an authorised purpose but involves a departure from the scheme (as with the linkage application). Once we escape the clutches of the tort model and think about the relationship in constitutional terms, the process for new authorisation might or might not involve going back to some or all of the participants. The mini-constitution, of course, should always be read as precluding purposes or projects that are contrary to the public interest; but, for those purposes that are judged to be compatible with the public interest, the constitution can provide mechanisms for authorisation that respect the participants but without necessarily requiring each and every participant to give a fresh consent or personal endorsement.

Secondly, what should we make of the situation where the mini-constitution guarantees that access will not be given to the police or for any purpose other than health-related research (for example, for insurance or employment purposes)? Here, there will be a question about whether this particular expression of private governance is consistent with the larger public interest. While it might be difficult to articulate a public interest in access that benefits private insurers or employers, the public interest in the prevention and detection of crime can scarcely be gainsaid. If such a tension is seen as one between private governance and the public interest, then the latter will prevail. However, if the tension is re-characterised as being between different strands of the public interest (the promotion of health-related research and the prevention and detection of crime), then a judgment will need to be made about where the balance of public interest lies.

Thirdly, should access be granted to assist with the identification of the victims of a disaster? Although the question of accessing biobanks in such circumstances is still relatively unexplored, there has been much more discussion about creating exceptions to the usual restrictions imposed by privacy and data protection laws (see Reidenberg et al. 2013). In response to emergencies, special measures have been adopted in some countries—for example, in the wake of the 2002 Bali bombing and the 2004 tsunami, the Australian government amended its privacy laws to permit the collection, use and disclosure of personal information where (in the context of such disasters) an emergency declaration is made.[14] Approving such initiatives, Joel Reidenberg, Robert Gellman, Jamela Debelak, Adam Elewa, and Nancy Liu have argued that (Reidenberg et al. 2013, 6):

> sharing information about missing persons is a legitimate objective in emergency situations, that data protection laws should accommodate this objective, and that…emergency circumstances require special exceptions to privacy rules that are proportional to the circumstances, including appropriate safeguards, and that remain in place only as long as the emergency circumstances necessitate.

If privacy and data protection (which, after all, will be regarded by many as fundamental rights) should accommodate such a pressing need, then should not a similar accommodation be made in respect of access to biobank data? On the one hand, the argument for granting access is that the purpose relates to a strand of the public interest—indeed, New Zealand's Assistant Privacy Commissioner, Blair Stewart, has put this in terms of acts that are "essential in the cause of common humanity" (Reidenberg et al. 2013, 1); on the other hand, the argument for denying access is that access for this kind of application has not been authorised by the participants (assuming that no implicit authorisation can be read into the mini-constitution). Although the purpose of this test-case application is neither health-related nor for research, the shape of the analysis is as in the first test-case: that is, a public interest reason (here, a public interest in identifying the dead) will exclude the terms of the private scheme unless it is set against a stronger public interest consideration (such as the public interest in maintaining voluntary participation in health-related research projects). However, given the evolving practice and philosophy in relation to privacy, together with the Swedish biobanking precedent, it seems likely that this would be one case where the public interest in access would be judged much stronger than the public interest in denying it.

Once again, of course, the tension might be resolved directly by the mini-constitution. For example, the mini-constitution might provide for flexible handling of applications that involve a non-authorised purpose that is, at least arguably, in the public interest (as with the identification of victims of disasters). If the mini-constitution sets out a process for dealing with applications of this class, then it can be set in motion without violating the consents of the participants. To a certain extent, participants put their trust in these processes. However, it is not a trust that is unconditional or unqualified: even if access decisions are properly made (that is to say, they are made in accordance with the prescribed process), where participants

---

[14] Formally, this was achieved by the introduction of Part VIa into the Privacy Act, 1988; see Reidenberg et al. 2013, 11–14.

are unhappy with the outcome, then they (like members of a club or association) retain the option of withdrawal (the nature of which should be specified in the mini-constitution).

14.5 Conclusion

Big Biobanks, coupled with the proliferation of genetic sequencing, seem to have an important place in future health-related research initiatives. This is, some think, the new paradigm. Some such biobanks already exist and, for them, there are particular problems to the extent that they are operating with governance frameworks that were elaborated before we came to see them as constitutions; and, even if we now view them as mini-constitutions, they probably will not provide for the kind of procedures that give the kind of flexibility that is required to deal with new approaches to feedback, unforeseen applications, and the like. For those Big Biobanks that are yet to be set up, there are many challenges but there is also an opportunity to be smart by writing constitutions that, while doing full justice to the interests of participants, articulate flexible and fair procedures for responding to a constantly changing backcloth of new technological and social developments.

References

Bell, J. 1993. Public interest: Policy or principle? In *Law and the public interest*, ed. R. Brownsword. Stuttgart: Franz Steiner Verlag.
Beskow, L.M., E.E. Namey, R.J. Cadigan, T. Brazg, J. Crouch, G.E. Henderson, et al. 2011. Research Participants' perspectives on genotype-driven research recruitment. *Journal of Empirical Research on Human Research Ethics* 6(4): 3–20.
Beyleveld, D. 2006. Conceptualising privacy in relation to medical research values. In *First do no harm*, ed. S.A.M. McLean, 151–163. Aldershot: Ashgate.
Beyleveld, D., and R. Brownsword. 2007. *Consent in the law*. Oxford: Hart.
Beyleveld, D., and R. Brownsword. 2015. Research participants and the right to be informed. In *Inspiring a medico-legal revolution: Essays in honour of Sheila McLean*, ed. P. Ferguson and G. Laurie. Aldershot: Ashgate.
Bovenberg, J., T. Meulenkamp, E. Smets, and S. Gevers. 2009. *Always expect the unexpected: Legal and social aspects of reporting biobank research results to individual research participants*. Nijmegen: Radboud University, Centre for Society and Genomics.
Brownsword, R. (ed.). 1993. *Law and the public interest*. Stuttgart: Franz Steiner.
Brownsword, R. 2004. The cult of consent: Fixation and fallacy. *Kings College Law Journal* 15(2): 223–252.
Brownsword, R. 2007. The ancillary care responsibilities of researchers: Reasonable but not great expectations. *The Journal of Law, Medicine & Ethics* 35(4): 679–691.
Brownsword, R. 2008. *Rights, regulation and the technological revolution*. Oxford: Oxford University Press.
Brownsword, R. 2009. Rights, responsibility and stewardship: Beyond consent. In *The governance of genetic information: Who decides?* ed. H. Widdows and C. Mullen, 99–125. Cambridge: Cambridge University Press.

Brownsword, R. 2012a. Guidelines for our genomic futures. *Rivista di Medicina* 20: 179–188.
Brownsword, R. 2012b. Informed consent in the information society. *Health Sociology Review* XIn3(special issue): 179–206
Brownsword, R. 2012c. Regulating brain imaging: Questions of privacy and informed consent. In *I know what you are thinking: Brain imaging and mental privacy*, ed. Edwards, S.J.L, Richmond, S, and G. Rees. Oxford: Oxford University Press.
Brownsword, R. 2013a. Crimes against humanity, simple crime, and human dignity. In *Humanity across international law and biolaw*, ed. B. van Beers, L. Corrias, and W. Werner. Cambridge: Cambridge University Press.
Brownsword, R. 2013b. Human dignity, human rights, and simply trying to do the right thing. In *Understanding human dignity*, ed. C. McCrudden, 470–490. Oxford: Oxford University Press.
Brownsword, R. 2013c. Regulating biobanks: Another triple bottom line. In *Comparative issues in the governance of research biobanks*, ed. G. Pascuzzi, U. Izzo, and M. Macilotti, 41–62. Heidelberg: Springer.
Brownsword, R., and M. Goodwin. 2012. *Law and the technologies of the twenty-first century.* Cambridge: Cambridge University Press.
Capps, B. 2013. Defining variables of access to UK biobank: The public interest and the public good. *Law, Innovative Technology* 5(1): 113–139.
Düwell, M., J. Braavig, R. Brownsword, and D. Mieth. 2014. *The Cambridge handbook of human dignity*. Cambridge: Cambridge University Press.
European Commission (EC). 2012a. *Biobanks for Europe: A challenge for governance.* Brussels: European Union.
European Commission (EC). 2012b. *Regulation of the European parliament and of the council on the protection of individuals with regard to the processing of personal data and on the free movement of such data.* Brussels: European Parliament.
Forgó, N., R. Kollek, M. Arning, T. Kruegel, and I. Petersen. 2010. *Ethical and legal requirements for transnational genetic research.* Berlin/Oxford: CH Beck with Hart.
Forsberg, J.S., and S. Soini. 2014. A big step for Finnish biobanking. *Nature Reviews Genetics* 15(1): 6.
Gliwa, C., and B.E. Berkman. 2013. Do researchers have an obligation to actively look for genetic incidental findings? *American Journal of Bioethics* 13: 32–42.
Heeney, C., and M. Parker. 2012. Ethics and the governance of biobanks: Understanding the interplay between law and practice. In *Governing biobanks*, ed. J. Kaye, S.M.C. Gibbons, C. Heeney, M. Parker, and A. Smart, 282–301. Oxford: Hart.
Human Genomics Strategy Group (HGSG). 2012. *Building on our inheritance: Genomic technology in healthcare.* London: Human Genomics Strategy Group.
Johnston, C., and J. Kaye. 2004. Does the UK biobank have a legal obligation to feedback individual findings to participants? *Medical Law Review* 12(3): 239–267.
Kaye, J., E.A. Whitley, N. Kanellopoulou, S. Creese, K. Hughes, and D. Lund. 2011. Dynamic consent: A solution to a perennial problem? *BMJ* 343: d6900–d6900.
Kaye, J., M. Hurles, H. Griffin, J. Grewal, M. Bobrow, N. Timpson, et al. 2014. Managing clinically significant findings in research: The UK10K example. *European Journal of Human Genetics* 22(9): 1100–1104.
Knoppers, B.M., M. Deschênes, M.H. Zawati, and A.M. Tassé. 2013. Population studies: Return of research results and incidental findings policy statement. *European Journal of Human Genetics* 21(3): 245–247.
Lunshof, J.E., G.M. Church, and B. Prainsack. 2014. Raw personal data: Providing access. *Science* 343(6169): 373–374.
Macilotti, M. 2013. Informed consent and research biobanks: A challenge in three dimensions. In *Comparative issues in the governance of research biobanks*, ed. G. Pascuzzi, U. Izzo, and M. Macilotti, 143–163. Heidelberg: Springer.
Milne, A.J.M. 1993. The public interest, political controversy, and the judges. In *Law and the public interest*, ed. R. Brownsword. Stuttgart: Franz Steiner Verlag.

Nuffield Council on Bioethics (NCB). 2011. *Human bodies: Donation for medicine and research*. London: Nuffield Council on Bioethics.

Personal Genome Project: United Kingdom (PGP-UK). 2014. Informed consent for enrolment in the PGP-UK. http://www.personalgenomes.org/static/docs/uk/PGP-UK_FullConsent_06Jun13_with_amend.pdf. Accessed 15 Feb 2015.

Presidential Commission for the Study of Bioethical Issues (PCSBI). 2013. *Anticipate and communicate—Ethical management of incidental and secondary findings in the clinical, research, and direct-to-consumer contexts*. Washington, DC: PCSBI.

Reidenberg, J.R., R. Gellman, J. Debelak, A. Elewa, and N. Liu. 2013. *Privacy and missing persons after natural disasters*. Washington, DC/New York: Center on Law and Information Policy at Fordham Law School and Woodrow Wilson International Center for Scholars.

Richardson, H.S. 2012. *Moral entanglements: The ancillary-care obligations of medical researchers*. New York: Oxford University Press.

Sample, I. 2013. *Critics urge caution as UK genome project hunts for volunteers*. The Guardian, 2013 November 7. http://www.theguardian.com/science/2013/nov/07/personal-genome-project-uk-launch. Accessed 15 Feb 2015.

Soini, S. 2013. Finland on a road towards a modern legal biobanking infrastructure. *European Journal of Health Law* 20(3): 289–294.

Steinsbekk, K.S., B. Kåre Myskja, and B. Solberg. 2013. Broad consent versus dynamic consent in biobank research: Is passive participation an ethical problem? *European Journal of Human Genetics* 21(9): 897–902.

UK Biobank. 2007. Ethics and governance framework of UK biobank. http://www.ukbiobank.ac.uk/wp-content/uploads/2011/05/EGF20082.pdf. Accessed 15 Feb 2015.

UNESCO. 2005. Universal declaration on bioethics and human rights. http://portal.unesco.org/en/ev.php-URL_ID=31058&URL_DO=DO_TOPIC&URL_SECTION=201.html. Accessed 15 Feb 2015.

Wellcome Trust and Medical Research Council (WTMRC 2012). 2012. *Assessing public attitudes to health related findings in research*. London: Opinion Leader.

Wolf, S.M., B.N. Crock, B. Van Ness, F. Lawrenz, J.P. Kahn, L.M. Beskow, et al. 2012. Managing incidental findings and research results in genomic research involving biobanks and archived data sets. *Genetics in Medicine* 14(4): 361–384.

World Medical Association (WMA). 2013. Declaration of Helsinki: Ethical principles for medical research involving human subjects. 64th WMA General Assembly; Fortaleza, Brazil. http://www.wma.net/en/30publications/10policies/b3. Accessed 15 Feb 2015.

Chapter 15
Ethical Dimensions of Dynamic Consent in Data-Intense Biomedical Research—Paradigm Shift, or Red Herring?

Bettina Schmietow

Abstract This chapter describes the rise of digital, personalised and adaptive forms of consent to use of biomaterial and/or data in research as a reaction to the limitations of traditional informed consent standards, which have been widely perceived as inadequate in large-scale biomedical studies. It uses the approach of a "dynamic consent" as an example, initially in particular in biobanks and genomics, in which participant and patient choices are to become more influential, and, in fact, central. In elaborating on some criticism to this approach and its proposed merits, it is argued that these forms of consent have potential for adapting data-intense research to new research requirements, but that the suggestion of "participant-centrism" as leading to a shift in research ethics would require more attention to ethical and social issues that have not yet been well developed. Preliminary anchoring points for such a shift are in understanding the redefinitions of participant-patient "autonomy" and "privacy" in view of the emergence of research, as well as social norms of data sharing, and in developing the link to a broader project of "citizen science".

15.1 Introduction

An intense debate has occupied bioethical scholarship for the past two decades, which aimed to analyse whether a perceived 'traditional' form of informed consent can and should adapt to emerging forms of research, from genomics and biobanking to increasingly virtual, global research networks assisted by online, openly shared genomic databases. Wider-scale, even "open" forms of consent where participants grant broad, unrestricted access to as yet unspecified future uses of their samples and associated data have been proposed. These approaches to safeguard the ethical requirements of informed research consent seem, however, to re-define the

---

B. Schmietow (✉)
Nuffield Council on Bioethics, 28 Bedford Square, WC1B 3JS London, UK
e-mail: b.schmietow@gmail.com

role of patient-participant autonomy and privacy, and focus instead on accountable governance approaches (Angrist 2009; Green and Guyer 2011; Friend and Norman 2013; Lunshof et al. 2008; Vayena and Tasioulas 2013). This chapter will examine to what extent a related – but still seemingly distinct – proposal to transform consent into a more dynamic process is positioned within the older discourse on the adaptability of informed consent procedures.

While these approaches are now employed in various contexts, and are framed within the language of customary consent requirements and attached values, the ethical implications of a particular "adaptive tendency" of informed consent remain vague and as yet under-explored. This might be the case because consent as a mechanism of research participant protection has proven normatively resistant, that is, it can also usefully, almost unrecognisably, standardise research governance and facilitate the active engagement of participants in generating massive, widely shared datasets. Indeed, "participant-centric" models are suggested to lead to a paradigm shift in research ethics towards more equal relationships between researchers and participants (Kaye et al. 2012; Vayena et al. 2013).

On the other hand, it seems to be indicative of a trend of consent in current 'bio-governance' that tends to disregard potentially problematic issues relating to digitised genomic research based on wide data sharing. It can thus serve as a kind of red herring, at least unless accompanied by incorporating further clarifying work on conceptual and practical limitations of informed consent, as well as the governance context in which these assumptions are played out.

Various versions of such digital consents and consent updates exist.[1] This chapter looks at the discourse around one proposed consent model—"dynamic consent"—that commentators argue is well-adapted to web-based data collection and storage. In the following, the idea and its ethical framing are summarised. The next part outlines and confronts criticisms that have been levelled against such an approach. These claim that dynamic consent is unduly individualistic and paternalistic in comparison to a broad consent model, and might be too demanding for potential research participants to be functional.

It will be suggested that these criticisms tend to be misguided or rely on controversial assumptions about the role of individual autonomy and other interests in research participation, and a strong presupposition that the research they enable is *per se* highly socially valuable. More constructively though, they can also be taken to point to potentially substantial ethical disagreements against the backdrop of a changed, digitalised form of research consent, in which research and a variety of other data uses are becoming increasingly blurred.

Reflections to this effect are offered by a more sociologically and politically informed approach on research ethics, which relates these developments to, for

---

[1] For example, patient platforms that could be employed for various purposes, linked to genome testing such as 23andMe, health data sharing combined with social networking such as Patients LikeMe; genomic information sharing platforms and tools such as Portable Legal Consent (see Kuehn 2013), and applications envisaged for medical record data sharing (see Dixon et al. 2014; Wee 2013; Wee et al. 2013).

example, the theory and practice of *citizen science*, a term popularised initially by Alan Irwin's work in the 1990s, when he referred to the cooperation between researchers and lay people in the context of developing solutions to more clearly emerging environmental threats (cf. Prainsack 2014). A similar trend has thus far only been alluded to in the bioethical debate.

In conclusion, it will be suggested that research ethics now finds itself at an ethical crossroad that would profit from being made more explicit, in that the reasons for the use of a approach such as dynamic consent extend beyond the concern of a tension between public and private interests in biobank-based genomics. Taking the blurring of responsibilities, as well as the possibilities for laypeople to engage in this research seriously, provides an opportunity for more nuanced ethical analysis. From this ethical perspective, then, the normative primacy of an individualist, human rights-based bioethics is put into question (Knoppers et al. 2014; Vayena et al. 2013), as is an appeal to constrain these through, e.g., genomic solidarity.[2] Moreover, "dynamic" approaches to biomedical research and consent express and enact newly emerging norms of sharing biological data, accompanied by an ambivalent commitment to sharing also research knowledge, expertise and power.

15.2 Consent and Biobank-Based Biomedical Research

Problems in the application of informed consent in biobank-based research have long been recognised: failure to take into account genetic (and beyond) connectedness, an inability to anticipate research uses, and insecurity if the right to withdraw from research can be respected. Since biobanks and their networks are 'research platforms' rather than specific, time-limited projects, factors of uncertainty of use, involvement of other parties and in relation to the necessary risk-benefit assessment abound (Shickle 2006). As technology is advancing towards global research networks, these seem likely to persist and only increase.

Many biobanks and virtual research repositories have adopted "broad consents"—ranging from relatively concrete forms, where participants accept or reject specific future uses and then are not recontacted, to "open" consent where participants agree to their samples or data being used for research projects with any purpose and/or receiving any type of funding, and/or in recognition of the fact that data security might be limited (Scott et al. 2012). Broad consent, more generally, has been considered an efficient, and yet ethically defensible or even pragmatically and morally preferable way to obtain consent by many commentators.

More specifically, some propose that broader forms of consent can bridge individualistic and solidarity-related aspects of data-intense, not bodily invasive research, such as in genomics (e.g. Hansson et al. 2006). Others, however, have strongly argued against this view, expressing concern that such a use will undermine "real" consent and exploit or at least disrespect participants (e.g. Hofmann 2009).

---

[2] Cf. Knoppers et al. 2014.

Clearly, the concept of informed consent remains the legal and conceptual cornerstone of current research ethics, with ethical and legal interpretations of its meaning and function filling volumes of scholarship. Some central fault lines in the debate on broad consent relate to the question whether consent should "by default" aspire to engage moral values such as individual autonomy and principles such as "respect for persons" (or at least not contradict these), or be sought rather insofar it has an impact on a person's objective, usually physical welfare.

In particular, the role of the guiding value of individual "autonomy" in data-intensive research remains, however, opaque. If someone donates a blood sample to an internationally linked biobank project or a saliva sample to a platform project such as or similar to 23andMe or the Personal Genome Project, how exactly does the signing of a consent form engage the person as an individual with particular morally relevant preferences, which will often be summed up as expressing respect for his or her "autonomy"? It certainly would offer the individual in question a certain degree of choice, although not all choice menus might be sufficient to respect a person's "autonomous" aspirations. An answer to this question would require a specification of the concept and value of autonomy of a certain generality, and yet,: does it really matter if consent processes facilitate participants' autonomous ends, particularly if the relevant research would often seem to pose negligible risks and burdens for the participant? The question remains contentious since even if the research performed does not impinge on a person directly and at this point in time, it might have morally relevant import at a later stage, for example when a study reveals as yet unknown medical conditions that affect the individual, but also his or her relatives. Very briefly, does consent cover the "dynamics" of linked and future uses, and in which sense does and should a one-off consent protect a person's "autonomy" and "privacy" (Hofmann 2004)?

According to this exposition, the "consent issue" may be perceived as raising central questions about the relationship between a person's biological and genetic material and his or her moral values, even personal identity, to which the digital and genomic turn in research only adds further complexity (cf. Boddington 2012). In one view, digitised consent approaches might simply replicate known ethical and conceptual concerns, for example, in having to assume a trusted patient-doctor relationship in the original context in which informed consent has been conceived. In another, a digital research environment might take on a "life of its own" from which one is not only unable to completely withdraw one's 'bio-input', but that in addition leads to a redefinition of the right to privacy, confidentiality and interests in personal autonomy in an equally contextual, adaptive way (cf. Nissenbaum 2010; Vayena et al. 2013).

As a consequence, this debate, as first applied to tissue and data (bio) banks, finds itself in a curious theoretical space. While it has been claimed that recent broad and digitised forms of consent hollow out the ethical core of consent (e.g. Hofmann 2009), at the same time these new practices appear to be highly adaptable, whichmight leave the impression that the central achievements of bioethics and human subject protection in research have been well preserved (cf. Elger 2010; Vayena et al. 2013).

Not the least of which, however, some important normative flanking considerations in research ethics rather than the value or reference interpretation of autonomy are being adjusted. This primarily concerns the long-established idea that, as per default, research is an exceptional activity in which the individual is entitled to stringent protections and his or her interests are considered to be of prime interest (cf. Helgesson and Eriksson 2008).[3] Newer models, treat research as an enterprise that should become a normal part of clinical treatment (Faden et al. 2013, 2014; Larson 2013). Overall, there is a movement in many countries to make available, and link, data from various sources beyond research. The data protection principle of data minimisation or data reduction to help protect privacy has partly yielded to a "open data" and "open access" movements, encouraged by research funders, and at least in part also sustained by patients and the public itself (Kaye 2012; Vayena et al. 2013).

15.3 Dynamic Consent and Personalised, Integrated Healthcare Research

The proposal of a "dynamic" or indeed "open" consent begins with acknowledging that research governance is out of sync with the unprecedented opportunities for data accumulation and sharing that advances in digital technology bring to the fore (Kaye 2011; Kaye et al. 2015, 1f.). This trend is seen to enable an entirely new model of medical research, which emerges "from a confluence of social media, citizen science, crowd-sourcing, and greater patient control over personal health information" (O'Connor 2013, 471).[4] Dan O'Connor refers to it as an "apomediated world" and thus, *apomediated research*. Apomediation "is envisioned as a more horizontal, peer-to-peer style of information exchange in which no single apomediary is essential to the process"(ibd.). In contrast to traditional hierarchical medical research, "apomediated research [...] is research in which information about the protocol—for example, its design and conduct—is apomediated, peer-to-peer, between individuals who may appear as both subjects and researchers"(ibd.). Crucially, apomediated research recognised as such would blur the distinction between the roles of health professional, researcher and lay person and their associated ethical responsibilities. As a consequence, protectionist regulation of research might seem both a categorical mistake and redundant, and indeed increase research inefficiency: "If regulations are there to protect subjects from researchers, what are

---

[3] Also, a tendency of depersonalisation or "datafication" has been observed, see e.g. Majumder 2005.

[4] Although the percentage of this research remains unquantified for the moment, it seems justifiable to claim that there is increasing attention to the possibilities of apomediated (in particular biomedical) research, citizen science and public and patient involvement (PPI). Patients and the public can become involved at various levels, from designing research protocols to organisational as well as ethics and governance issues. An example could be decisions concerning which therapy a trial tests and if it uses an active or placebo comparator. Other aspects would be considered more "operational" research participation. The British Medical Journal, for instance, is committing to the "patient revolution in healthcare" by implementing, among other things, patient peer review. See Richards and Godlee 2014; Welsman et al. 2014.

regulations for when subject and research seem to be one and the same?" (O'Connor 2013, 471).

In contrast with paper-based, one-off consent, the dynamic approach to consent mirrors the platform-like character of new research forms, using social media types of communication interfaces. These allow for tailoring consent to a wider variety of research initiatives, in a more open and more flexible manner. Participants could also be approached on a case-by-case basis for emerging projects (Kaye et al. 2015, 2). As Wee suggests, "dynamic consent has evolved in step with changing technological developments over the past decade. It can be described as encompassing a range of characteristics that enable interactive ways for individuals to express and change their consent virtually immediately, at any time, and on a continuous or ongoing basis" (Wee et al. 2013, 344).

At the outset, dynamic consent and similar initiatives had been proposed as a "technique of alignment" between patient concerns and research needs, with the ambition that these could "transform the debate from questions of public good versus individual autonomy, and cost versus practicality to one where the concerns of the patient are aligned with the needs of medical research" (Kanellopoulou et al. 2011). In this first guise, the digitised, dynamic form of consent is a technological tool rather than the expression and enactment of particular concepts of patient autonomy or even an emerging 'bio-citizenship'. The patient or participant is free to engage with research and new projects if he or she so wishes, but might also remain passive and/or change her or his mind about this choice.

Simultaneous to abstaining from self-labelling and committing to a particular form of ethics,[5] it has been framed within a larger discourse of patient- and participant-centrism to "place patients at the centre of decision-making" and "understand and value the central role that patients have in research as the providers of information and biological material" (Kaye et al. 2015, 2, 5). Autonomous patient-participant choice as a prerequisite of consent is re-defined as expressing and leading to individual *empowerment*, which would involve more benefits than a value-neutral, digitised menu of choice. It would make the individual's choices better, more informed, research more robust, and, the whole research environment more trustworthy. It would also change the status of the participant into becoming a co-producer of the research outcome. In this second sense, the dynamic interaction, promoted by a culture of sharing, suggests an anti-paternalistic movement in which researchers and participants are equalised (O'Connor 2013).

As these quotations suggest, the ethical undertones of the proposal appear complex and ambivalent: Is dynamic consent merely a 'technique of alignment' or a transformative tool of ethics and governance to foster pro-actively individual autonomy and the accountability of science to the public in the age of digital biology, or even more than that? What is its relationship to the perceived 'traditional' consent?

Dynamic consent would at first sight seem to implement "real" consent that protects people's autonomy and privacy—broadly conceived—without significantly

---

[5] The approach is characterised as "an example of how IT can be used to satisfy the legal and regulatory requirements for research consent, while at the same time providing a personalised communication interface for interacting with patients, participants and citizens" (Kaye et al. 2015, 1).

impeding the pace of research. It may, in addition, even increase and improve recruitment possibilities and transparency of research uses. At this level, the demand for an alignment is indeed merely 'technical' and presented as being able to combine both ethics and practicality. If there are secure ways of storing and sharing people's bio-input, then the main ethical problem left is one of consent or dissent in the face of the choice of being a contributor, and against an assumption that indeed people want to share (Mamo et al. 2013, 921).[6] Nonetheless, the number and interlinking of bio- and data-hubs pose increasing risks that samples can be de-identified and assurances that participant details remain anonymous unrealistic. Commentators have proposed, however, that these risks can also be pro-actively minimised through, again, adapting consent to "accommodate the fluidity of data-flows in research networks" (Kaye et al. 2015, 2).

15.4 Criticisms of Dynamic Consent

Although the outlined model would seem timely—a logical consequence of current research developments and the infrastructures implemented—the proposal quickly triggered criticism. Steinsbekk et al., for example, have advanced arguments to the effect that broad and dynamic consent approaches appear to be decidedly distinct. They advocate the position that dynamic consent, though initially appealing, suffers from a number of problematic implications. Its normative baggage is seen to have significant potential to prove counter-productive in a complex and highly interconnected future research environment. Part of their criticism is related to empirical questions about the motivation and amenability of members of the public to involvement in research. These empirical concerns will be largely set aside here with the focus more directly on normative objections, although questions about the public's willingness to consent "dynamically" to research do to some extent factor in the argument.

Steinsbekk et al.'s main criticism appears to be that dynamic consent—rather than opening a wider field of choice, in the anticipation of a general public willingness to share—would foster an overly individualistic and by implication un-solidaristic approach to research governance (Steinsbekk et al. 2013, 901). Individuals would have to "always make an informed consent to both primary and secondary use of their data", independent of the rational justification and relevance of the additional information provided (ibid.,, 898). The authors' preferred broad consent model, in contrast, would seldom ask for re-consent, and when asking, this would be for important reasons—for example, in the context of research that is particularly con-

---

[6] A number of disclaimers that might significantly impact on any ethical conclusions apply: the knowledge of participants about secondary uses of donated research material tends to be limited, and people's theoretical concerns about privacy often do not match actual behavior. Conceptions of privacy expectations are framed around data security, which is widely seen as precarious; and preferences for sharing tend to be expressed on the condition of socially beneficial research.

troversial. The relevant (biobank) institution and/or research ethics committee will decide which situations are sufficiently important to warrant re-consent (Steinsbekk et al. 2013, 898).

Individuals might otherwise demand control and feedback over something to which they would not seem to have continuous and encompassing entitlements of such a kind, since it is information and body material that is not of immediate relevance for themselves: "Biomedical research […] is not primarily about our own health but rather about potential health benefits for future generations. An important reason for active engagement and participation in biomedical research is thereby lacking compared with general health care" (Steinsbekk et al. 2013, 900). Steinsbekk et al.'s claim is that "participant-centrism" and an implicit primacy of private interests therein are misguided, while the pragmatic approach to consent accounts sufficiently well for a donor's "autonomy", which appears to be the only, adequately clear ethical marker. Broad consent, as consent to governance rather than as requiring active opt-ins on a continuous basis (in the dynamic approach to which they refer in their comment), then appears to be a fundamentally different approach.

The proponents of dynamic consent, in Steinsbekk et al.'s reading, suggest instead that the bidirectional, ongoing, interactive process between patients [research participants] and researchers is to be morally preferred. passive participation would consequently appear to be "morally inferior or otherwise problematic" while they proposed the following focus:

> "The core of this debate, as we see it, is what it means to be 'adequately informed' and whether giving consent on broader premises is valid or not […],[as] 'more information' in itself does not necessarily make a consent more informed. Rather, it is relevant information that makes a consent informed." (Steinsbekk et al. 2013, 898f.)

This is highlighted despite the fact that these forms of consent would seem to lie on a continuum rather than being proper alternatives, since dynamic consent can be particularly broad (Kaye et al. 2015, 3). Indeed, it could become "open" if anonymity and "privacy" turn out to be illusory, and thus not desirable in terms of participant empowerment as currently cast; and this was acknowledged to the wider public as such.

As has been pointed out before, broad consent to research participation might be ethically justified as informed consent (e.g. Sheehan 2011). If consent is primarily about non-coerced choice, then the amount and quality of information communicated in any given healthcare or research context might be contingent on its power to enforce "autonomy", in other words: one might make "autonomous", for instance reflective, but still "ignorant" choices. This disregards the fact that consent-as-choice satisfies only some autonomy-reductionist, liberty-based approaches to consent in research ethics and governance that critics need to justify. Importantly, the autonomy debate might be unhelpful in any case since it usually focuses on individual autonomy, while digitised biomedical data-intense research is often aimed at aggregated information, at least initially.

As we cannot be sure that data and their anonymisation are secured and what impact an unintended data release might have, it is also not possible to simply assume that individual-level rights and interests such as a presumed respect for individual autonomy become irrelevant. However, concentrating on the guiding power and value of "autonomy", as well as relevant information and understanding of research, expressed through a standard of informed consent that can be stretched quite comfortably, does not seem to alleviate the conceptual and practical vacuum (cf. Hofmann 2004, 240). Even if broad consent might be acceptable to some or even most patient-participants, there is a danger that this approach might be misused or at least function as the initially mentioned red herring. This would mean that its promotion is strongly suggestive of individual control and understanding, while its framing as an extension of broad consent-to-governance might be cutting short further options of engaging with uncomfortable participant wishes and concerns (cf. Francis and Francis 2013; Hofmann 2004, 240).

Most importantly for this point, Steinsbekk et al. seem to be proposing that the gap in expertise between researchers and participants is real and should be upheld, first and foremost in the interest of an efficient science (Steinsbekk et al. 2013, 900). Biobank research is non-invasive, and "carries the potential for important medical breakthroughs and beneficial medical inventions", and so "it is morally problematic if consent procedures unnecessarily reduce or prevent these opportunities" (Steinsbekk et al. 2013, 901). While the authors acknowledge that "a true democratic and participatory model of medical research in general would be a model wherein citizens were allowed to impact which kind of research initiatives they thought would have the biggest effect on promoting health and reducing the burdens of disease in a society", they negate any necessity of questioning "today's framing of biobank research" without further argument (Steinsbekk et al. 2013, 900).[7]

This leads to the problem of how "deflationary" in terms of *research expertise* research governance today should be, and thus, to the viability of a new project of *citizen science*. Both questions do not relate directly to the topic of consent as it has primarily been discussed in bioethics. The wider context of how the normative superstructure of consent—if broad, dynamic, open, etc.—merges into research governance, has usually been discounted. Although both a more pro-active, dynamic consent and a more ethically neutrally broad consent capture real features of the current developments—one more focused on bottom-up governance aspects that highlight participant interests as central, the other placing more emphasis on top-down governance suggestive of 'genetic or genomic solidarity'—the solitary application of the consent mechanism, quite independent of its content, means that the wider and future-oriented impact on participants and public of this type of research might remain very much understated.[8]

---

[7] For biobank research and similar big data projects suggesting increasing secondary uses, the tension is again in the currently unclear importance of large data sets and analyses, with the parallel affirmation that the effects will be transformative. Consent is then overburdened on various levels, as data subjects are asked to consent to "research that is not research" (Ioannidis 2013), as well as the considerations on "apomediated research" above.

[8] Both uses for the "public good" (leading to an argument for "data citizenship") and individual objections to unethical or unjust use are the dimensions that are of importance here (Francis and Francis 2013).

As far as consent 'in itself' is concerned and rules, research governance appears to be merely a matter of individual choice—even if either more altruistic or more control-focused, highlighting expert information delivery, understanding and uptake in standardisable fashion, and also as generally conservative: a post-hoc agreement, in that there is no conceptual space or accompanying mechanism to incorporate the social, political and cultural research context, or to anticipate future developments.

Despite the fact that consent might remain an important means to promote individually valuable autonomy in its various shades, and as an expression of respect for persons, the context in which such consent is obtained and ethically contemplated has changed. In particular, research and healthcare, public and private research organisations and funding are increasingly interlinked (O'Connor 2013; Vayena et al. 2013). Steinsbekk et al.'s arguments try to resist this "apomediated world" in their critique of dynamic consent.

The dynamic consent approach described here is, on the contrary, an expression of the growing emphasis on less hierarchical relationships between researchers and participants in the life sciences and digitised research environments, although its proponents have not committed to a more explicit re-framing of research ethics as research governance of this kind. Other emerging ethical values pointing beyond consent, that is, beyond a participant- or research-centrism, however, could be pinpointed by integrating the debate on citizen science and genetics, which are, however, beyond the scope of this paper.

15.5 Conclusion: Emerging Principles?

In the analysis presented above, the problems to which broad consent reacts are largely due to the rationale and organisation of biobank and data-intense biomedical research itself. Dynamic and other adapted consent approaches are mainly a consequent technological development that cannot by themselves solve any of these persisting and increasing ethical issues. They will also have to rely on technological solutions to be inclusive for a wider variety of users, and in addressing issues such as the digital divide (Kaye et al. 2015, 3). From both a conceptual and governance point of view, however, it is important to emphasise that technologically-driven approaches cannot redress the ambivalent role of both privacy protection and a simultaneous expectation of ever-growing data sharing that currently reigns in biomedical and genomic research, as well as its surrounding ethical discourse.

It has been suggested here that even if adaptive consents use a language of individual and public empowerment, this is done within broad terms of individual liberty and autonomy that are, as yet, virtually neutral on the normative desirability and possibility of levelling the playing field between researchers, clinicians, and the public, and between different individuals and different publics. Consequently, there is scope to engage with the ongoing discourse on patient empowerment and citizen science that might counteract the curious adaptive tendency of a research ethics framed almost entirely around consent. Engaging with these more inclusive and interdisciplinary movements towards patient empowerment and citizen science may help to expand research ethics' attention to issues such as participants as co-producers of research knowledge and that go beyond questions of valid and informed consent to research.

"Citizen science" can be defined as a conceptual tool and practical movement that reacts to a "need for scientists and members of the public to cooperate in the face of complex societal challenges" (Prainsack 2014). The move to open and digitised research environments provides a link to the renewed question as to whether, and in which way, science should be a matter of civic engagement and participation. This seems particularly relevant as new norms and values are emerging in the blurred, open and digitised health-research biomedical complex.

These greatly complicate the technological and governance view that the dynamic approach represents, as ethics also moves from protectionist to becoming transitional and dynamic, perhaps leaving individually embodied interests behind. Barbara Prainsack summarises that 23andMe, for example, expresses the following normative stances (Prainsack 2014):

(1) that data access is an end in itself, and that is it not the role of the service provider, but of the end user, to decide on the utility of the data;
(2) that data is not something that flows only in one direction—from the service provider to the user—but also vice versa; and
(3) that in a system that relies on data contributions—and partly also contributions to data interpretation and analysis—from volunteers, the definition of expertise is changing.

The crucial normative point here seems to be that rather than the individual person (or formerly, patient) or the freedom and value of research for a common or even private good, but data—their accumulation, interlinking and sharing—have gained normative primacy. Dynamic consent, though aiming to facilitate patient engagement, is currently limited to enabling big data research, which is a paradigmatic shift driven by the availability of technology. Participant-centrism that is similarly technology-focused might be unable to ameliorate and flexibilise this situation should the interests of participants remain locked within the "formality of consent" (Hofmann 2004), if these broader but primary considerations are only to be decided further down technology advance.

Acknowledgement The author, research officer at the Nuffield Council on Bioethics, UK, would like to express her thanks to the European School of Molecular Medicine, Milan, Italy and the HeLEX Centre, University of Oxford, UK, where parts of this contribution were conceived.

References

Angrist, M. 2009. Eyes wide open: The personal genome project, citizen science and veracity in informed consent. *Personalized Medicine* 6(6): 691–699.
Boddington, P. 2012. *Ethical challenges in genomics research*. Berlin/Heidelberg: Springer.
Dixon, W.G., K. Spencer, H. Williams, C. Sanders, D. Lund, E.A. Whitley, et al. 2014. A dynamic model of patient consent to sharing of medical record data. *BMJ* 34: g1294.

Elger, B. 2010. *Ethical issues of human genetic databases: A challenge to classical health research ethics?* Farnham: Ashgate.

Faden, R.R., N.E. Kass, S.N. Goodman, P. Pronovost, S. Tunis, and T.L. Beauchamp. 2013. An ethics framework for a learning health care system: A departure from traditional research ethics and clinical ethics. *Hastings Center Report* 43(s1): S16–S27.

Faden, R.R., T.L. Beauchamp, and N.E. Kass. 2014. Informed consent, comparative effectiveness, and learning healthcare. *New England Journal of Medicine* 370(8): 766–768.

Francis, L.P., and J.G. Francis. 2013. Data citizenship and informed consent. *American Journal of Bioethics* 13(4): 38–39.

Friend, S.H., and T.C. Norman. 2013. Metcalfe's law and the biology information commons. *Nature Biotechnology* 31(4): 297–303.

Green, E.D., and M.S. Guyer. 2011. National human genome research institute. Charting a course for genomic medicine from base pairs to bedside. *Nature* 470: 204–213.

Hansson, M.G., J. Dillner, C.R. Bartram, J.A. Carlson, and G. Helgesson. 2006. Should donors be allowed to give broad consent to future biobank research? *Lancet Oncology* 7: 266–269.

Helgesson, G., and S. Eriksson. 2008. Against the principle that the individual shall have priority over science. *Journal of Medical Ethics* 34: 54–56.

Hofmann, B. 2004. Do biobanks promote paternalism? On the loss of autonomy in the quest for individual independence. In *Blood and data: Ethical, legal and social aspects of human genetic databases*, ed. G. Árnason, S. Nordal, and V. Árnason, 237–242. Reykjavik: The Centre for Ethics and University of Iceland Press.

Hofmann, B. 2009. Broadening consent – And diluting ethics? *Journal of Medical Ethics* 35: 125–129.

Ioannidis, J.P.A. 2013. Informed consent, big data, and the oxymoron of research that is not research. *American Journal of Bioethics* 13(4): 40–42.

Kanellopoulou, N.K., J. Kaye, E.A. Whitley, S. Creese, D. Lund, and K. Hughes. 2011. Dynamic consent – A solution to a perennial problem? *BMJ* 343: d6900.

Kaye, J. 2011. From single biobanks to international networks: Developing E-Governance. *Human Genetics* 130(3): 377–382.

Kaye, J. 2012. The tension between data sharing and the protection of privacy in genomics research. *Annual Review of Genomics and Human Genetics* 13: 415–431.

Kaye, J., L. Curren, N. Anderson, K. Edwards, S.M. Fullerton, N. Kanellopoulou, et al. 2012. From patients to partners: Participant-centric initiatives in biomedical research. *Nature Reviews Genetics* 13(5): 371–376.

Kaye, J., E.A. Whitley, D. Lund, M. Morrison, H. Teare, and K. Melham. 2015. Dynamic consent – A patient interface for 21st century research networks. *European Journal of Human Genetics* 23: 141–146.

Knoppers, B.M., J.R. Harris, I. Budin-Ljøsne, and E.S. Dove. 2014. A human rights approach to an international code of conduct for genomic and clinical data sharing. *Human Genetics* 133(7): 895–903.

Kuehn, B.M. 2013. Groups experiment with digital tools for patient consent. *JAMA* 310(7): 678–680.

Larson, E.B. 2013. Building trust in the power of "Big Data" research to serve the public good. *JAMA* 309(23): 2443–2444.

Lunshof, J.E., R. Chadwick, D.B. Vorhaus, and G. Church. 2008. From genetic privacy to open consent. *Nature Reviews Genetics* 9: 406–411.

Majumder, M.A. 2005. Cyberbanks and other virtual research repositories. *The Journal of Law, Medicine & Ethics* 33(1): 31–39.

Mamo, L.A., D.K. Browe, H.C. Logan, and K.K. Kim. 2013. Patient informed governance of distributed research networks: Results and discussion from six patient focus groups. *AMIA Annual Symposium Proceedings* 2013: 920–929.

Nissenbaum, H. 2010. *Privacy in context. Technology, policy, and the integrity of social life.* Stanford: Stanford University Press.

O'Connor, D. 2013. The apomediated world: Regulating research when social media has changed research. *The Journal of Law, Medicine & Ethics* 41(2): 470–483.

Prainsack, B. 2014. Understanding participation: The "citizen science" of genetics. In *Genetics as social practice*, ed. B. Prainsack, S. Schicktanz, and G. Werner-Felmayer, 147–164. Farnham: Ashgate.

Richards, T., and F. Godlee. 2014. The BMJ's own patient journey. *BMJ* 348: g3726.

Scott, C.T., T. Caulfield, E. Borgelt, and J. Illes. 2012. Personal medicine – The new banking crisis. *Nature Biotechnology* 30(2): 141–147.

Sheehan, M. 2011. Can broad consent be informed consent? *Public Health Ethics* 4(3): 226–235.

Shickle, D. 2006. The consent problem within DNA biobanks. *Studies in History and Philosophy of Biological and Biomedical Sciences* 37: 503–519.

Steinsbekk, K.S., B.K. Myskja, and B. Solberg. 2013. Broad consent versus dynamic consent in biobank research: Is passive participation an ethical problem? *European Journal of Human Genetics* 21: 897–902.

Vayena, E., and J. Tasioulas. 2013. Adapting standards: Ethical oversight of participant-led health research. *PLoS Medicine* 10(3): e10001402.

Vayena, E., A. Mastroianni, and J. Kahn. 2013. Caught in the web: Informed consent for online health research. *Science Translational Medicine* 5: 173fs6.

Wee, R. 2013. Dynamic consent in the digital age of biology. *Journal of Primary Health Care* 5(3): 259–261.

Wee, R., M. Henaghan, and I. Winship. 2013. Dynamic consent in the digital age of biology: Online initiatives and regulatory considerations. *Journal of Primary Health Care* 5(4): 341–347.

Welsman, J., A. Gibson, J. Heaton, and N. Britten. 2014. Involving patients and the public in healthcare operational research. *BMJ* 349: g4903.

Chapter 16
Using Patent Law to Enforce Ethical Standards: Proposal of a New Patent Requirement

Jan-Ole Reichardt

Abstract Clinical trials are important instruments for achieving scientific progress within the life sciences. However, while they are of the utmost importance to our translational efforts, they are also highly expensive. To save costs, they are often relocated into developing countries where the protection of study participants is minimal. Such relocation is not necessarily amoral, as those in charge might nevertheless adhere to high ethical standards. However, relocation is problematic if it entails the exploitation of vulnerable participants. How can such exploitation and violation of ethical standards within the life sciences be prevented? Adopting a pragmatic approach to research ethics, this paper suggests using the incentivising mechanisms of our patenting process to tackle the challenge of the prevailing unethical treatment of human subjects in life science research. By linking the granting of economic benefits via patents to the fulfilment of ethical requirements, the paper makes an important contribution to the question of how "ethical excellence" can be achieved in one of the most lucrative areas of global research.

16.1 Introduction: Reversing the "Race to the Bottom" Effect

In our partially globalised world, it is the power to transfer financial resources and goods that has been internationalised the most. As a matter of fact, companies are widely free to establish and relocate the linking segments of their commodity chains and service infrastructures—on a worldwide scale and in accordance with their own preferences. This also applies to research and development services in the area of health care product development. This area is a vast global business estimated to reach total revenues of nearly USD 1.3 trillion by 2018, which represents an increase

---

J.-O. Reichardt (✉)
Institute for Ethics, History, and Philosophy of Medicine, University of Münster,
Von-Esmarch-Straße 62, 48149 Münster, Germany
e-mail: Jan-Ole.Reichardt@uni-muenster.de

of about 30% over the 2013 level (IMS 2014, 1). What is unique about this kind of enterprise is the necessity to perform extensive clinical trials that leave large footprints in the cost accounting. Studying drug effects on hundreds or even thousands of trial participants is a highly expensive endeavour, even if the developmental costs were a good deal below the self-proclaimed one billion Euro per newly approved drug (Interpharma 2014), as Light and Warburton (2011) argue convincingly.

As long as the companies' dominant preference lies in optimising the financial gains and efficiency of their investments, it is economically reasonable to cut costs by relocating clinical trials into the lowest-cost regions available. To identify those regions, several factors are taken into account, such as cost-efficient infrastructure, wage levels, availability of subsidies or additional opportunities to externalise costs due to, for example, lax environmental standards. From the investor's perspective, regulatory measures (e.g. worker protection, safety precautions) that diminish the return on investments are regarded as cost-driving factors and therefore economic annoyances. As long as primarily economic factors are taken into account, regions with fewer and lower cost-driving demands offer competitive advantages over those regions with more extensive requirements. The almost global range of these companies in relocating their research facilities, when combined with the fact that the transfer of knowledge is—in contrast to physical goods—not subject to tariffs and trade barriers, leads to a "race to the bottom" effect (Tabb 2003) with regard to ethical standards. This effect refers to a continuous erosion of quality standards that is caused by competing players who are thereby aiming to increase their relative competitive advantages. In this way, clinical trials are moved to countries where the least is demanded from those who are offering themselves as "human guinea pigs".

According to Glickman et al. (2009, 816) there has been an annual growth of 15% in the number of active FDA (Food and Drug Administration)—regulated investigators working outside the US since 2002, while the number of US-based investigators decreased by 5.5%. And, as Miller (2011) notes by referring to the US Department of Health and Human Services, the number of foreign trials for US drugs over the past two decades has shown a 2000% increase, leading to a situation in which, by 2008, approximately 80% of drug applications approved by the FDA contained data from foreign clinical trials. The relocation of clinical trials to increase economic efficiency is not necessarily amoral as those in charge might still adhere to high ethical standards. However, relocations become ethically problematic if they entail or even rely on the exploitation of vulnerable participants. Several studies have documented the widespread occurrence of ethical fraud in relation to clinical trials (see, e.g., SOMO 2008, Carome 2014). The independent, not-for-profit research and network organisation SOMO (2008, 3) identified the lack of voluntary, informed participation and proper consent as the most common issues, while the most alarming examples of misconduct were tests with experimental drugs of which the safety for testing in humans had not yet been fully established.

In pharmaceutical research, human trial participants are—apart from animal subjects—the most vulnerable group involved in the translational process. In contexts in which poverty compels people to accept even the worst offers, people are particularly vulnerable to exploitation and therefore in serious need of regulatory protection.[1] How can such exploitation of poverty and violation of ethical standards within the life sciences be prevented? Adopting a pragmatic approach to research ethics, this paper suggests making use of the incentivising mechanisms of our patenting processes to tackle the challenge of prevailing unethical treatment of human subjects in life science research. It is argued here that the granting of economic benefits through patents should be linked to the fulfilment of ethical requirements in order to achieve ethical excellence.

This approach has two crucial advantages. First, it aims to establish a regulatory procedure that does not leave the protected worse off than they were without our well-meant interventions. As long as research facilities can be moved easily, we need a solution that avoids the foreseeable migration of clinical trials into still unregulated regions. Thus, an ethical standard has to be obligatory regardless of the regional context of the research. Second, the approach relies on the fact that almost all research-conducting institutions are interested in the commercial use of their findings and strive for a patent-based market monopoly to do so. This protection, which is usually seen as being granted to honour and encourage a researcher's contribution to the welfare of the awarding community, should be extended to require compliance with basic ethical standards as well. While the concrete ethical requirements are debatable, it is advisable to start with the ethical principles of the WMA Declaration of Helsinki (2013). Non-compliance with these ethical principles as well as any misconduct that can be attributed to a neglect of the duty of supervision should be regarded as undermining any eligibility to claim patent registration. The remainder of the paper is organised as follows: first, an analysis of what patents are, is provided; second, a strategy for how patenting could be enhanced ethically is developed and lastly, four objections to the recommended approach are discussed.

16.2 Patenting: A Cultural Invention

The starting point of the pragmatic approach adopted in this paper is the nearly analytical truth that if social activities are exclusively aimed at optimising economic prospects, all investments are evaluated in light of their contribution to this single value agenda. This implies, for instance, that all cost-reducing strategies helping to reach an economically promising future should indeed be realised. Where- and whenever the relocation of a specific part of a value-producing network appears economically sound (and manageable), there is a dominant reason to do so. And wherever a strategy of "abiding to the law" seems economically less promising than a strategy of "pay your penalties, but only when caught and all defensive measures fail", the latter will be the barely hidden recommendation from a purely economic

---

[1] For an overview of research with vulnerable participants and compelling arguments for a sound regulation of such endeavours, see Siep (2014).

standpoint. Of course, people tend to have more than just economic preferences. However, with regard to the management of larger holdings, where investor-investment relations are of a merely remote nature, economic motives predominant and non-economic outcomes only evaluated in terms of their economic contributions, ethical considerations have to pay off or they will probably be disregarded. This is also true for commercially motivated research that usually patents findings to secure some monopoly rights of commercial use.

If we—as a society—have an interest in additional values besides economic ones, a pragmatic way to achieve this is to establish regulatory frameworks that render strategies to externalise costs inapplicable while transforming strategies of acquiescing according to the rules into economically favourable ones. "Pragmatic" in this sense means to take certain realities of our current world as given when we simply lack the power to change them and to try to improve these conditions by acting from within the system. Thus, it is not the aim of the argument presented here to change the system itself, by, for example, trying to alter people's and companies' behaviour based on ethical insights and motivation. This does not imply, however, considering every aspect of the current system as ethically acceptable, as will be discussed here later. The key idea behind the patent-approach is to convert ethical behaviour into an economically efficient and worthwhile business. Ethically ignorant behaviour is a costly attitude and a form of economic mismanagement. As a means of governmentally provided market protection, patents can be used as ideal mechanisms to introduce ethical considerations into our regulatory frameworks.

The topic of patenting and the question of what defines a patent, can be approached from different perspectives and with various interests in mind. From a lawyer's perspective, a patent is primarily a monopoly-providing title, which grants its holder a negative right: namely, the right to prohibit the unauthorised production, use, offering and actual sale of the patented subject (Brougher 2013). These negative rights are a subset of what is commonly referred to as Intellectual Property Rights (or just IP) and as such are part of the same superset as trademarks, copyrights or industrial designs (Kur and Dreier 2013). From a sociologist's perspective, Brougher's definition is an adequate description of a legislative practice. This practice relies on a concept of patentability that is based on a specific set of ideas and cultural inventions.[2] From a historian's perspective, the status quo is just the current state of a long-term evolution to which many causal factors have contributed during its multi-centennial course of development, and that is still dynamic and likely to change again in the future.[3] From a moral philosopher's perspective, these different sets of ideas that form the ideological basis of specific patent legislation, their normative implications as well as our argumentative resources to justify or criticise them and to promote or contest their recognition, are of key interest.

---

[2] For an overview of different dimensions of property rights systems and their variations over time with regard to the transformation of intellectual property rights, see Carruthers and Ariovich (2004).

[3] For a profound analysis of these developments, see May and Sell (2006), May (2007), Adams (2009) and Sherman (2013).

Epistemologists have taught us that if there was only one true idea of patentability—one that had to be found and could not be designed in accordance with our needs (like a law of nature)—we would need to conduct research to find out what patents are and how the most adequate concept of patentability would look.[4] However, truth is the wrong criterion by which to measure the quality of a cultural invention. Rather, our own preferences are back in the game to form the one and only relevant quality criterion. Hence, the status quo of our patenting processes is not written in stone but subject to social determination and negotiation. Patents can thus be considered to be cultural tools, and, as with all tools, their overall quality (or usefulness) is observer-specific and depends on the extent to which they support their observers' objectives. As different observers tend to have diverse preferences, these will result in assorted objectives and consequently varying quality judgments. Therefore, it is no surprise that various authorities have welcomed and introduced dissimilar concepts of patentability into their respective legislations.

Simultaneously, there are attempts to harmonise existing patent laws. The most industrialised countries in particular are promoting the global adoption of their preferred understanding and handling of IP. In spite of the extensive actions of the United Nation's World Intellectual Property Organization (WIPO), founded in 1967, the World Trade Organization (WTO)—especially with its 1994 Trade-Related Aspects of Intellectual Property Rights (TRIPS) agreement—,[5] and more region-specific harmonising activities,[6] the world-wide harmonisation of IP legislations is still incomplete. So, despite the standards that have been widely established by the TRIPS agreement, we are still in a situation where disparate legal IP frameworks exist. These frameworks require different things and are offering varying levels of protection while being geographically limited to the jurisdiction of the acknowledging authorities. Current legal patent frameworks typically differ in what they (i) acknowledge as patentable, (ii) regard as requirements for their granting of a patent, (iii) grant as entitlements by doing so (and for how long) and (iv) as to what they refer for justificatory purposes.

This regulatory diversity provides different incentives leading to various motivational forces and actions of those in charge. So, which IP systems should be targeted for implementing the suggested ethical standards? Again, from a pragmatic point of view and in order to achieve what is currently the broadest possible coverage, it makes sense to focus on the most influential and widespread IP systems first, that is, those of US and European origin and the most recent WTO agreements on IP. This

---

[4] Such an hypothesis could be formulated by proponents of a meta-ethical position of moral realism, which—by doing so—could suggest that specific ideas of patenting came closer to the "truth" of what is actually owed each to the other.

[5] For a descriptive account of WTO IP measures see Niemann (2008), and for an IP law commentary, Cottier and Véron (2011).

[6] For regional examples of this global harmonisation trend see Shi (2008) on China and the EU, Malhotra (2010) on TRIPS in India and for the intra-European harmonisation process see Seville (2009), Hugenholtz (2013) and Geiger (2013).

does not preclude, however, that the suggested ethical complement of patent law can and should also be applied to additional patent systems as well.

16.3 Ethical Enhancement of IP Systems

Looking at today's most influential IP systems, we find them unmotivated to introduce ethical incentives into business administration: there simply are no ethical requirements with which an inventor must comply in order to secure a patent. In fact, patent eligibility requirements are completely removed from the social and environmental conditions under which the respective research took place (Storz 2014, 1):

> Patentability requirements in the two major patent jurisdictions [are] novelty, non-obviousness/inventive step, enablement/written description, best mode, and sufficiency of disclosure.

But ethical requirements are not only untapped within the status quo, they are also absent from regulatory debates, although public opinion commands some veto powers to exclude sufficiently contested objects (like human-derived products) from EU-patentability (EPC 2010, Art. 53a):

> European patents shall not be granted in respect of: (a) inventions the commercial exploitation of which would be contrary to 'ordre public' or morality; such exploitation shall not be deemed to be so contrary merely because it is prohibited by law or regulation in some or all of the Contracting States[.]

However, morality has only been introduced into our patent procedures with regard to the question of the kinds of things that should or should not be patentable (Moufang 2008). Apart from this paragraph, morality has not been incorporated into EU, US and WTO patent law, nor elsewhere and has not been related to the very conditions under which the research has taken place. In this vein, Thambisetty concludes that our patenting processes are widely neglectful of the bioethical aspects that go beyond those of mere patentability (2007, 247):

> The relationship between law and morality is particularly fraught in the sphere of patent law. There is reluctance to concede that morality and patentability intersect, and a number of legal scholars have argued that patent law was not intended to encompass moral or ethical judgments on inventions. Although it is possible to historically trace moral concerns within legal doctrine, for example in the controversy over patenting playing cards in the 19th century, and more recently over the protection of contraceptives, patent systems in Europe and the United States remain largely unreceptive to bioethics.

This situation should be changed and given that—contrary to the knowledge that will be gained by research—patents are, as argued above, not intrinsic components of a research achievement, this situation *can* be changed (at least in theory). The rather simple idea suggested here is to include an additional obligation of "ethical excellence" to the group of patentability requirements with which the inventor must already comply in order to earn whatever entitlement the patent confers.

But even if some ethical requirements were implemented into *our* patenting systems, we would still face the challenge that these systems have a limited regional scope. Companies could insist that they have already "over-complied" with the ethical standards set by, for example, the Philippine government for the Philippine region. Such a region-relative introduction of ethical requirements would fail to protect the most vulnerable. It would only contribute to the research costs that are subject to extensive cost cutting efforts and would quickly fall victim to the aforementioned "race to the bottom" with regard to regulatory hurdles. Therefore, a further patch needs to be added to the project of using patent law to achieve ethical excellence: the granting of a patent must ask the respective applicant to meet the ethical requirements of the granting body regardless of the regional context where the actual research took place. If a patent candidate wanted to be granted monopoly rights in Europe, s/he would have to comply with European standards of ethical excellence even if the research itself was delegated to a research contractor in Bangladesh. The idea behind this *cuius regio, eius religio* approach to patenting is to maintain the legality of carrying out research in inexpensive contexts in order to avoid worsening the situation of those who already live in precarious conditions, while economically discouraging undue exploitation of individual vulnerabilities and of statutory loopholes.

There are two key advantages of this proposition. First, it converts the race to the bottom into a race to the top effect as it incentivises compliance with the highest standards to gain access to the most promising commercial markets. This would also take the pressure off industrialised countries to participate in this race to the bottom by eroding their own standards. By doing both, it would establish a regulatory procedure that does not leave the protected worse off than they were without our well-meant interventions. Second, the approach does not rely on the goodwill of or voluntary commitment to social responsibility but combines existing institutions with the currently prevailing market logic to strive for higher ethical standards. It relies exclusively on the fact that almost all research-conducting institutions are interested in the commercial application of their findings and aim for a patent-based market monopoly to do so in the most promising way.

If such an approach to improve current patenting systems and best practice standards in clinical research were to be adapted, the remaining theoretical task would be to provide an interpretation of ethical excellence that seems adequate for various research contexts. What should the concrete ethical requirements be? The main challenges in this process are the diversity of research and its contexts on the one hand and the array of people's beliefs about moral requirements (in general and in particular contexts) on the other. While it is unlikely to achieve universal approval for suggested ethical standards, it helps to start with some ethical core components that can command the backing of the majority. For that reason, and in favour of an immediate implementation of ethical requirements within the life sciences, we should start with a set of standards that already enjoys majority support. Here, the recently revised WMA Declaration of Helsinki (2013) for a specific regulation of research involving human subjects and the more general UN Universal Declaration of Human Rights (1948) offer appropriate standards that should be regarded as the starting ethical minimum.[7]

Further work on the specifics of these recommendations is of course required, but success in implementing them in one or another form could contribute significantly to a globalisation of which we as human beings had more reason to be proud. To have more than a merely symbolic impact, it would be important to substantiate those requirements of ethical excellence and misconduct so they can be operationalised, reviewed and assessed in a legal framework. That might precipitate controversial debates, but that is also true of "sufficiency" regarding the minimum level of invention in scientific excellence and—to start with—a provisional working basis, which would be acceptable.

Given that the current patenting systems consider questions of morality only with respect to patentability, we have plenty of leeway to introduce further moral requirements without causing interference with what we already have. Nevertheless, one difficulty remains: the challenge of convincing a sufficiently large group of politically active stakeholders to regard these recommendations as ideologically well-fitted, suitable, as well as consistent with and welcome additions to the status quo of patent regulation. The following section will discuss four key objections that could be raised against the idea of combining patent requirements with ethical standards.

16.4 Discussion of Objections

This paper aims to initiate a change of international IP systems, which so far has received little attention and which can face several practical difficulties as well as ethical objections. To assess the chances of its implementation, substantial information on the respective status quo is needed. This can easily become a Sisyphean task, as the analysis of national and transnational patent law is both complex and dynamic, thus making it difficult to develop and maintain the necessary levels of expertise. The good news is, however, that much less heroism is required, because at second glance, we do not need to be experts on the whole to improve a small part of it. In the following section, four main objections to the idea of combining patent law with ethical requirements will be discussed that have been or might be raised by legal and philosophical scholars. These are first, the argument that patent law was not intended to encompass moral or ethical judgments on inventions; second, the issue of information on (un)intended consequences of patent processes; third, the demur that the

---

[7] This declaration asks, for example, for the informed consent (§ 26) of every trial participant, requiring adequate information about all relevant aspects of the study, including aims; methods; sources of funding; possible conflicts of interest; institutional affiliations of the researcher; anticipated benefits and potential risks of the study; the discomfort it may entail; post-study provisions; and the right to refuse to participate in the study or to withdraw consent to participate at any time without reprisal.

patent system is immoral to a degree that it should rather be rejected as a whole; and fourth, the question of how capable we actually are with regard to modifications and ethical improvements of IP systems.

16.4.1 Patent Law Was Not Intended to Encompass Moral or Ethical Judgments on Inventions

As Thambisetty (2007, 247) notes, a significant number of legal scholars have argued that patent law and ethical judgement are two distinct areas of competence systematically and should be kept separate. This objection is based on at least two misleading premises: (i) that patenting law as it is has no normative or moral implications and (ii) that the introduction of ethical considerations into patenting law would degrade and encumber an otherwise systematically distinct area of legislation. The first premise can easily be rebutted, as the granting of additional monopoly rights is already a dramatic deviation from normative reservation, in that it imposes new obligations on previously unfettered market participants and merges scientific accomplishments with financial prospects.

As far as the second premise is concerned, an already normative concept that grants merit-based monopoly rights will not lose the coherence of its inner logic if additional requirements for merit are raised. Moreover, the attitude of regarding *no* ethical requirement at all as sufficient for the granting of exclusive exploitation rights can be considered as an ethical position of its own instead of being ethically agnostic with respect to patent law. Furthermore, since proponents of patent law are already forced to acknowledge their deviation from normative reservation and the merging of a scientific logic with economic aspects, the recommendation to introduce further ethical requirements into the patenting process faces no fundamental objection and is thus a legitimate candidate for further discussion.

16.4.2 Unbiased Information Is Difficult to Come by, Which May Not Matter

When the consequences of a given regulatory approach are to be judged, and when some of its parts are to be modified, it would seem that a great deal of research must be done first in order to draw valid conclusions about the given system—perhaps more than one has the possibility to conduct within a reasonable amount of time. To elaborate on these consequences, it is quite helpful to distinguish between different types of such consequences: (i) those that were intended by a specific stakeholder group, (ii) those that were individually presumed to follow from a specific regulation, (iii) those that were publicly declared as ensuing and (iv) those that actually did ensue. If, for example, an individual's accountability for specific consequences or someone's integrity is to be investigated, the true motives and presumptions of those in charge are of utmost importance. But since this paper focuses on policies and not on authors and their motives, what is relevant here are the actual consequences of a specific legislation. Yet, while we are interested in the actual consequences, we can only provide individual assumptions about those consequences—although some might be more justified than others and what we are seeking are those assumptions that are the most and best justified. However, the best justification would require insights into alternative world histories (what would or would not have happened if we had other laws) and those alternative world histories are—by design—merely notional and notoriously difficult to test. Moreover, the second best justification—one that draws on other countries' experience with different legislations—might also be out of reach, since it is oftentimes difficult to find real world examples of sufficiently similar collectives running sufficiently dissimilar IP systems—or one might just encounter too much interference with and from uncontrolled events to get clean data. In all those cases, a more theory-driven approach might prevail and we will have to be content with merely well-founded estimates.

When we are finally provided with the respective models and hypotheses, the matter becomes even more complicated: since the alleged consequences of IP systems are subjects of ideological warfare, we face a challenge of potentially biased discourse participants and publications. This is not surprising, because IP systems have the potential to dramatically alter capital flows and power structures within societies. Consequently, the stakes are high and some stakeholders are unafraid to deny facts, assert falsehoods and refuse even the most obvious conclusions to obtain strategic advantages with regard to their own political agenda.

Without any intention to judge a specific author's individual bias, it is apparent that some of them have less empirical evidence available to substantiate their claims and prefer the recitation of ideological narratives when asked about the specific impact of our current patent processes. Take, for instance, Brougher's statement about the exclusive capabilities of monopoly rights (2013, v):

> Development of medical technology [...] would not be possible without the promise of a monopoly period that allows companies to recover the development costs of a product, make a profit, and also finance the development of future products.

This rather bold non-existence claim regarding alternative avenues of scientific progress is simply wrong as a quick glance at the history of science and technology demonstrates: historically, medical technology was developed long before IP systems were introduced and medical progress has also been achieved by publicly funded non-profit organisations such as universities. It might be difficult to meet the economic challenges of today's life science research, but universities are still successful in the development of medical technologies and—if they were remunerated differently—could also conduct the more expensive phases of clinical research. If, however, some ideological premises were accepted as inescapable certainties (e.g., for-profit research organisations within the framework of a pure market economy), one might regard some assertions as far less peculiar than they look from a less confusing position. Scanning through the literature of economics, such untenable assumptions and a rather neo-liberal mindset are quite common.

At the same time, anthropologists with a stronger focus on aspects of individual livelihood security have identified and stressed the negative side-effects of IP systems on the poor and most vulnerable inhabitants of less industrialised nations. Based on her research on development politics, Lanoszka (2003, 194) concludes: "The leading industrialized countries must pay attention to the social and economic needs of developing countries. However, this would likely require a considerable departure from the existing attitude towards IPRs derived from western legal practice and now institutionalized in the TRIPS Agreement". Similarly, Menghaney (2013, 7013) concludes that, regarding her experience with patent-produced hurdles impeding access to medical treatment, "patents continue to obstruct access for people with HIV to drugs to treat neglected co-infections, such as drug resistant tuberculosis and hepatitis C. The need for governments to find ways to tackle patent barriers to safeguard public health is as relevant now as it was a decade ago". So if we want to discuss the ethical aspects of IP systems, it seems necessary to adopt a transnational perspective, to delve deeply into socio-economic research, to obtain context-specific and unbiased information, and to avoid succumbing to someone's ideological line before we can finally deliver a comprehensive and exhaustive assessment of a specific patent process.

Luckily, this would only be necessary if we needed such an exhaustive assessment of patent procedures at all. But we do not—for the following reason: one simply does not have to know all the consequences of a specific legislation to improve it in a particular area. Instead, it is perfectly acceptable to argue in favour of an only relative and not absolute preferability and to develop aspect-based improvements.

16.4.3 The Reasonability of Improving What Might Nevertheless Remain Harmful

If we regard the patenting system as morally defective and only try to improve some of its aspects, this does not guarantee that our efforts might ever result in something that is—as a whole—morally acceptable or even desirable. Should we not choose a more radical approach then, refuse to meddle with what is presumably beyond repair and simply call for an abolition of the status quo? This rather idealistic objection can be complemented with the accusation of "conservative ideology". The use of IP systems to promote ethical values, it could be claimed, might undermine the political resistance to patenting and could be criticised as "ethical greenwashing".

How can we respond to this objection? Here again, a pragmatic argument can be advanced. Whenever we lack the capacity to achieve ethically superior results, it is reasonable to improve at least that of which we are capable, even if some harmful legislation can at best be mitigated. Can it nevertheless be reasonable to demand to abolish legislation without any hope of being successful? Yes, this can be the case, because it is no less reasonable to conceptualise our institutional ideals and to think about the laws we would introduce or eliminate if we only had the respective political authority. Doing so and focusing in particular on our existing IP systems, one could indeed wonder if they would not qualify as irretrievably disastrous institutions that might never reach the thresholds of at least a moral break-even point.

At the same time, one might also try to envision a more helpful interpretation of the whole IP concept. Therefore, even if a specific interpretation and implementation of IP were actually harmful, we do not have to discard the idea itself as inevitably morally deficient. Gold (2013, 193) argues in this direction, when he asks us to reject the view of an intrinsically amoral patent system and engrain morality into the idea itself:

> Instead of accepting that patent law is irresponsible and in need of discipline, as the subjugation approach does, or of confusing one set of rights (patents) with another (human rights) as in both the subjugation and integrated approaches, we need to expand our expectations of patent law from simply increasing wealth to increasing well-being.

This provides a good argument for the approach to introduce moral requirements into our IP systems developed in this paper. Of course, focusing on the potential benefits of an idea might still lead us to the conclusion that it might never produce enough net benefit to be embraced and therefore does not invalidate opposition in principle to IP systems on the theoretical level. Furthermore, it does not invalidate such an opposition on the practical level either—at least for as long as we regard our chances to remove the patenting system as sufficiently promising, which leads us to the last objection: our capacity to change the system.

16.4.4 Political Capacities or the Basis on Which a Strategy of Incremental Change Could Be More Pragmatic

It might be time to recall the yet unanswered question as to how capable we actually are with regard to modifications and ethical improvements of IP systems. When it comes to regulatory ideals, it is a fully comprehensible position to radically oppose IP systems. However, when we switch from normative theory to the actual implementation of our normative ideas, a revolutionary approach will have more extensive success requirements than a rather moderate approach of incremental modifications. Furthermore, it seems almost impossible to succeed with a revolutionary agenda when international agreements are affected. For that reason, Thambisetty (2007) recommends that a strategy of incremental improvement be applied in preference to revolutionary renewal, which would better correspond to our law's tradition of a step-wise evolution (2007, 266f.):

> Complex institutional relationships within the patent system make it difficult to implement changes to patentability rules and doctrine that are not directly related to or adapted from past experience. [T]he incremental advance of the rules react uneasily to 'policy overhaul' type of arguments that are often required when debating ethical implications of unprece-

dented subject matter as in the case of biological material in the early years of the biotechnology revolution. The long-term projection of this process of change is further complicated by institutional 'stickiness' that can make it hard to reverse undesirable or simply inaccurate interpretations of the rules. The virtual inevitability of incrementalism may come as a disappointment to some idealists. However the appeal of incrementalism in policy formulation in general and in the patent system in particular is high because [an] 'overhaul' type of reform introduces formidable legal and political risk. 'Satisficing', rather than goal maximizing, is the preferred criterion and slight improvement compared to past performance is favored.

In addition to this well-considered advice, it seems that a revolutionary approach would not only challenge the institutional logics of legislative evolution, but could also provoke a much stronger resistance from those stakeholders who are interested in a continuous existence of the status quo. For that reason alone, political considerations are indispensable, if the best expected value with regard to ethical improvements is to be realised.

16.5 Conclusion

The primary objective of this paper has been to disincentivise the wilful acceptance and advancement of external costs within the contexts of bioscientific research. To do so, it has recommended a regulatory amendment that would render cost externalisation strategies inapplicable, would transform a strategy of acquiescing to the rules into an economically favourable one, and ethically ignorant behaviour into a costly attitude and a simple form of mismanagement. The best option to reach these goals has been located within the context of IP systems, where ethical requirements are—without formal or substantive reasons—actually underrepresented and could be added without further ado and without fear of inadvertent inconsistencies. Although such an endeavour would also and inevitably spawn stakeholders who—with good reason—would fear losses and therefore insist on a perpetuation of the status quo, we should try to overcome whatever obstacle they might actually produce. Additionally, an approach of just incremental improvements seems to produce small collaterals at most, so all arguments advocate a movement in the envisioned direction. As long as we as society are granting patents to foster progress and honour accomplishments, we should not refrain from asking for *ethical excellence* in return.[8]

---

[8] The argument of this paper—to introduce an additional requirement of "ethical excellence" into our patenting processes—has greatly benefited from discussions with many people. I would like to thank those who provided invaluable feedback on different occasions, in particular the members of the BMBF research group *Research Ethics—Current Challenges in Preclinical, Clinical and Public Health Research* as well as Sarah Chan, John Harris, David Hunter, Marcel Mertz, Thomas Pogge, Heiner Raspe, Catherine Rhodes, Annette Rid, Sarah Ruth Sippel and Daniel Strech.

References

Adams, J.N. 2009. History of the patent system. In *Patent law and theory*, ed. T. Takenaka, 101–131. Cheltenham: Edward Elgar Publishing.

Brougher, J. 2013. *Intellectual property and health technologies*. Berlin/Heidelberg/New York: Springer.

Carome, M. 2014. Unethical clinical trials still being conducted in developing countries. The world post. 2014 Mar 10. http://www.huffingtonpost.com/michael-carome-md/unethical-clinical-trials_b_5927660.html. Accessed 21 Feb 2015.

Carruthers, B.G., and L. Ariovich. 2004. The sociology of property rights. *Annual Review of Sociology* 30: 23–46.

Cottier, T., and P. Véron. 2011. *Concise international and European IP law*. Alphen aan den Rijn: Kluwer Law International.

European Patent Convention (EPC). 2010. Article 53. Exceptions of patentability. Amended by the EPC Revision Act of 29.11.2000. http://www.epo.org/law-practice/legal-texts/html/epc/2013/e/ar53.html. Accessed 21 Feb 2015.

Geiger, C. 2013. The construction of intellectual property in the European Union: Searching for coherence. In *Constructing European intellectual property*, ed. C. Geiger, 5–23. Cheltenham: Edward Elgar Publishing.

Glickman, S.W., J.G. McHutchison, E.D. Peterson, C.B. Cairns, R.A. Harrington, R.M. Califf, et al. 2009. Ethical and scientific implications of the globalization of clinical research. *The New England Journal of Medicine* 360(8): 816–823.

Gold, E.R. 2013. Patents and human rights. A heterodox analysis. *The Journal of Law, Medicine & Ethics* 41(1): 185–198.

Hugenholtz, P.B. 2013. The dynamics of harmonization of copyright at the European level. In *Constructing European intellectual property*, ed. C. Geiger, 273–291. Cheltenham: Edward Elgar Publishing.

IMS Institute for Healthcare Informatics (IMS). 2014. *Global outlook for medicines through 2018*. http://www.imshealth.com/portal/site/imshealth/menuitem.762a961826aad98f53c753c71ad8c22a/?vgnextoid=266e05267aea9410VgnVCM10000076192ca2RCRD&vgnextchannel=a64de5fda6370410VgnVCM10000076192ca2RCRD&vgnextfmt=default. Accessed 21 Feb 2015.

Interpharma. 2014. Forschung bei Entwicklung von Medikamenten [Research in drug development]. http://www.interpharma.ch/forschung/1805-forschung-bei-entwicklung-von--medikamenten. Accessed 21 Feb 2015.

Kur, A., and T. Dreier. 2013. *European intellectual property law*. Cheltenham: Edward Elgar Publishing.

Lanoszka, A. 2003. The global politics of intellectual property rights and pharmaceutical drug policies in developing countries. *International Political Science Review* 24(2): 181–197.

Light, D.W., and R. Warburton. 2011. Demythologizing the high costs of pharmaceutical research. *Biosocieties* 6(1): 34–50.

Malhotra, P. 2010. *Impact of TRIPS in India*. Basingstoke: Palgrave Macmillan.

May, C. 2007. *The world intellectual property organization. Resurgence and the development agenda*. Oxon/New York: Routledge.

May, C., and S.K. Sell. 2006. *Intellectual property rights. A critical history*. Boulder/London: Lynne Rienner Publishers.

Menghaney, L. 2013. Patent injustice: How India brought cheap HIV drugs to Africa. *BMJ* 347: f7013.

Miller, T. 2011. *"Explosive" growth in foreign drug testing raises ethical questions*. Interview with Arthur Caplan. PBS Newshour. 2011 Aug 23. http://www.pbs.org/newshour/rundown/sending-us-drug-research-overseas. Accessed 21 Feb 2015.

Moufang, R. 2008. Ethical requirements and limitations of patent protection for biotechnological inventions [Ethische Voraussetzungen und Grenzen des patentrechtlichen Schutzes biotechnologischer Erfindungen]. In *Intellectual property. Copyright or exploitation entitlement?*

[Geistiges Eigentum: Schutzrecht oder Ausbeutungstitel?], ed. Depenheuer, O., and K.N. Peifer, 89–109. Berlin/Heidelberg/New York: Springer.

Niemann, I. 2008. *Intellectual property under concurring treaty regimes – The relation of WIPO and WTO/TRIPS [Geistiges Eigentum in konkurrierenden völkerrechtlichen Vertragsordnungen].* Berlin/Heidelberg/New York: Springer.

Seville, C. 2009. *EU intellectual property law and policy.* Cheltenham: Edward Elgar Publishing.

Sherman, B. 2013. Towards a history of patent law. In *Intellectual property in common law and civil law*, ed. T. Toshiko, 3–15. Cheltenham: Edward Elgar Publishing.

Shi, W. 2008. *Intellectual property in the global trading system.* Berlin/Heidelberg/New York: Springer.

Siep, L. 2014. Ethical criteria for medical research in developing countries [Ethische Kriterien für medizinische Forschung in Entwicklungsländern]. In *Sisäisyys & Suunnistautuminen: Juhlakirja Jussi Kotkavirralle*, ed. A. Laitinen, J. Saarinen, H. Ikäheimo, P. Lyyra, and P. Niemi, 730–755.

Stichting Onderzoek Multinationale Ondernemingen (SOMO). 2008. SOMO briefing paper on ethics in clinical trials #1: Examples of unethical trials. http://somo.nl/publications-en/Publication_2534. Accessed 21 Feb 2015.

Storz, U. 2014. Patentability requirements of biotech patents. In *Biopatent law: European vs. US patent law*, ed. U. Storz, M. Quodbach, S.D. Marty, D. Constantine, and M. Parker, 1–21. Berlin/Heidelberg/New York: Springer.

Tabb, W.K. 2003. Race to the bottom? In *Implicating empire. Globalization & resistance in the 21st century world order*, ed. S. Aronowitz and H. Gautney, 151–158. New York: Basic Books.

Thambisetty, S. 2007. The institutional nature of the patent system. Implications for bioethical decision-making. In *Ethics and law of intellectual property. Current problems in politics, science and technology*, ed. C. Lenk, N. Hoppe, and R. Andorno, 247–267. Farnham: Ashgate.

United Nations (UN). 1948. Universal declaration of human rights. Resolution 217 A (III) adopted by the United Nations General Assembly on 10. December 1948. http://www.un.org/en/documents/udhr/index.shtml. Accessed 21 Feb 2015.

World Medical Association (WMA). 2013. Declaration of Helsinki: Ethical principles for medical research involving human subjects. 64th WMA General Assembly; 2013; Fortaleza, Brazil. http://www.wma.net/en/30publications/10policies/b3. Accessed 21 Feb 2015.

World Trade Organisation (WTO). 1994. TRIPS: Agreement on trade-related aspects of intellectual property rights. Annex 1C of the Marrakesh Agreement Establishing the World Trade Organization. Marrakesh, Morocco. http://www.wto.org/english/tratop_e/trips_e/t_agm0_e.htm. Accessed 21 Feb 2015.